Exercise ViralDuo
After-Action Report & Improvement Plan

North Florida Amateur Radio Club
Santa Fe Amateur Radio Society
Gainesville, Florida
February 2, 2019

Gordon L. Gibby MD

Copyright © 2019 Gordon L. Gibby MD KX4Z

All rights reserved, but emergency oriented amateur radio clubs may copy any portion or all for any reasonable non-profit purpose

ISBN: 9781796291995

Although the Author is responsible for the content of this work, the Draft was shared for comment and discussion with: W4UFL, KG4HBN, KI4IGX, AA3YB, KG4VWI, KM4JTE, W2SRP, AC4QS, KK4BFN, K1CE, W4OZK, W4JIR, K4MVR, KN4MQQ, and others. All input was taken seriously and appropriate changes made.

DEDICATION

This text is dedicated to all the leaders who made it possible to hold the 2019 Amateur Radio Emergency Communications Conference.

Team 2 hard at work!

Kevin Rulapaugh's MARC- Unit's Tower set up at Northside Park to assist amateur radio operators -- perfect example of interoperability training.

CONTENTS

Chapter		Page
	Acknowledgments	vii
1	Exercise ViralDuo Overview	1
2	Initial Tasks	5
3	Exercise Injects	11
4	Results	17
5	Exercise Categorization	27
6	Analysis of Core Capabilities	29
7	Improvement Plan	49
	Appendix: Exercise Written Evaluation Document	53
	Appendix: ICS-309 Documents Received	55
	Appendix: Quieting the Inverter Generator	63
	About the Conference	67

ACKNOWLEDGMENTS

So many people jumped in to volunteer to do so much for this conference, that even trying to list them is fraught with risk. But lets try! Jeff Capehart handled several talks, and listened to every crazy idea that I came up with and tried to make the ARES CONNECT work and created our fantastic Google sign-up page. Susan Halbert not only handled VHF antennas, she also took care of all the registration and money collection. Mike Ridlon (where do I start?) got us the ROOMS for the conference, headed up one of the Strike Teams, took care of ALL the audio visual and arrangements and details, and got us a new VHF gateway location right before the Conference, and held the finest "Soldering Session" our Club has ever done just a couple weeks before, in preparation for the soundcard building done at the Conference; Alvin Osmena volunteered to help with ANYTHING and was everywhere doing just that. Rosemary Jones not only got the food (!!) but also helped move things and took great photos. Leland Gallup took on one of the most difficult talks and also worked with Shannon Boal on the HF Antenna Talk. Shannon also taught how to help people pass ham exams! Joe Bassett stepped up to the plate to do the all-important LEADERSHIP talk and hit it out of the park! Karl Martin, with Ben Henley, handled the SEC end of things, gave me huge mentorship, and helped organize a far more effective "FARPOC" team that will do great things in the future! Scott Roberts got everyone up to speed on the incredibly important PIO task and also served ably in the net control slot during the Exercise. Joe Bassett kept his streak going with an absolutely fantastic voice traffic talk. Kevin Rulapaugh got the tallest tower many of us have ever seen right to Team Two and also did great demo's. Jeff Capehart taught and taught and taught the important basics of modern computer usage to bring us up to speed with working with professionals. James Lea gave people a first hand view of what being stuck at a shelter was really like when you weren't planning on tit! Vann Chesney handled the SAFETY aspect; and John Troupe was everywhere and also crimping power poles for everyone. Stewart Reissener – well, he made SIXTEEN simple inexpensive go-box frames that will put 16 more stations ready to go. Susan and Alvin got people building easy VHF antennas. Col. Huckstep, new to our group, twice managed to get me the most experienced Mentors possible within the County-- and even the nation – to keep me going in the right direction, and also carried load after load of gear; and Steve Waterman got me COML's to go over our exercise and give us great tips and mentorship (and he's the guy who got me into SHARES and also into go-boxes).

With ALL THESE GREAT PEOPLE working together, no wonder we have so much fun in Alachua County!

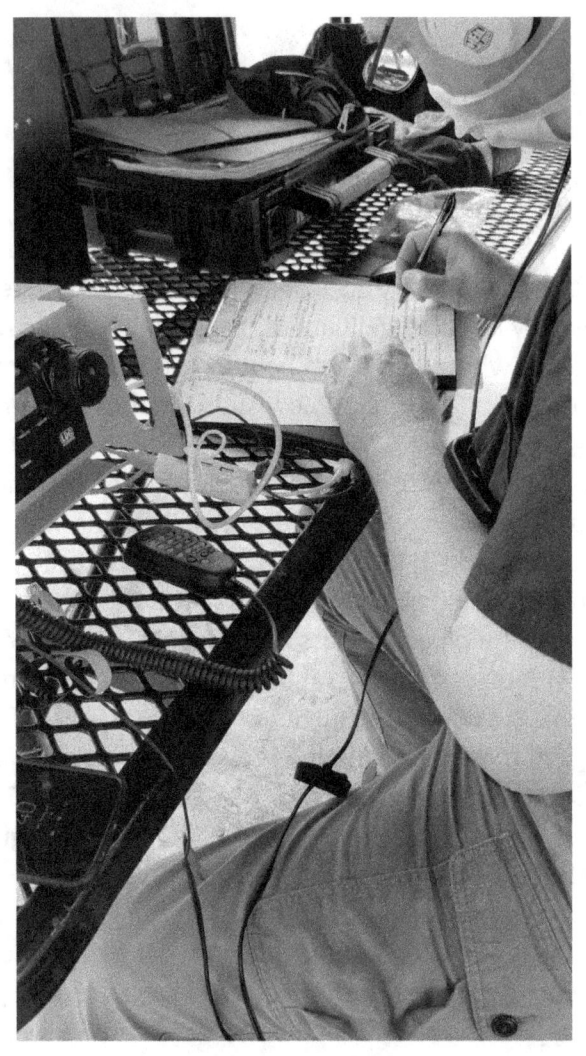

1 EXERCISE VIRAL DUO OVERVIEW

When the North Florida Amateur Radio Club and the Santa Fe Amateur Radio Society team up to hold an emergency communications Conference, there will ALWAYS be a deployment Exercise!

Exercise ViralDuo was the deployment exercise of the 2019 (Florida) Amateur Radio Emergency Communications Conference, held at Santa Fe College by the Santa Fe Amateur Radio Society and the North Florida Amateur Radio Club on Feb 2/3 2019. The Exercise was introduced in stages, with pre-conference training by way of multiple training emails, including an early release of the ICS-205 and 205A; a briefing for leaders the night before the Conference, and the availability of the entire ICS-201 by way of a Conference wifi web server beginning at 0700 on the day of the Exercise. The ICS-201 served as the participant manual for this Exercise.

The scenario and exercise were written in standard HSEEP format, with multiple injects using envelopes, as well as radio injects from 2 local amateurs and 2 distant amateurs, unknown to the participants of the Conference The Exercise was reviewed by two experienced COML's from other states, and by a former head of FEMA. As much as practicable, their suggestions were followed. The FEMA expert was concerned that the exercise as written was too complicated, so simplifying changes were made as much as possible.

EXERCISE DOCUMENTS

Document	URL
ICS-201	https://qsl.net/nf4rc/2019Conference/ICS201.pdf
ICS-205	https://qsl.net/nf4rc/2019/EmergencyConference/ICS-205.pdf
ICS-205A	https://qsl.net/nf4rc/2019/EmergencyConference/ICS-205A.pdf
ICS-206	https://qsl.net/nf4rc/2019/EmergencyConference/ICS206.pdf
Master Scenario Event List	https://qsl.net/nf4rc/2019Conference/ExerciseViralDuoMSEL.pdf
Compiled pre-conference training	https://qsl.net/nf4rc/2019/EmergencyConference/ PRECONFERENCEEMAILSCONCATENATED.pdf

TIMELINE AT THE CONFERENCE

At 0900 a briefing on ICS and this exercise commenced, with choices of the Team Leaders and Shelter Managers. The large crowd of roughly 55 participants were released at 0930 to travel to their deployment locations. There were 4 off-campus distractor participants pre-arranged and unknown to the conference participants, in three states. The Exercise had 1-hour "operational periods" and ended shortly after 1210.

At the introduction to the exercise, hard copy of the ICS-201, 205, 205A and 206 was given to both team leaders and to their management group, known as the "FARPOC". Additionally, all these documents were available on the Conference WIFI server, and could be captured as PDFs and stored by any of the Conference participants.

EXERCISE SCENARIO

On Jan 15, Dan F. and two of his neighbors went to Okeechobee County, Florida to hunt quail. They had a successful hunt and cleaned and consumed several quail and brought others home to their families. One day after returning, Dan and one neighbor fell ill with a severe pneumonia, soon followed by 5 other members of their families. Admitted to Biggs-Rinker Community Hospital, Dan and three others died within 48 hours; one victim survived after transfer to a tertiary hospital intensive care unit (intubated and placed in a drug-induced coma temporarily).

Polk County public health officials quickly quarantined all three families and known contacts, as CDC and Florida State Health Dept officials were flown in – **but two contacts evaded the quarantine, one headed south and one north.**. As 7 nurses and doctors of the community hospital began violent coughing, CDC officials announced a *novel, highly contagious and highly virulent form of avian influenza was responsible* and asked for drastic public health measures immediately given the 40+% fatality rate.

Unfortunately, new cases were quickly discovered centered around fast food restaurants along I-75, the suspected route of the escaping neighbors, and additional fatalities were soon recorded. The bodies of the two escaped neighbors were among the fatalities. With disease now spreading along both I-4 and I-75, hundreds of thousands of panicked Floridians began self-evacuating northward, causing Georgia and Alabama to quickly shutter their borders – but then new cases burgeoned in Valdosta, Waycross, GA, and Dothan Alabama. The nation went on the highest public health alert to the rapidly spreading and unusually fatal disease. Southeast states shut down interstate traffic at borders except for medical supplies. Trains are being scrubbed and inspected for stowaways at all Southern borders. Air travel is shut down in 16 States. Fuel was quickly becoming scarce in Florida, and there were concerns about dwindling food stores.

On January 30 with an estimated 18,000 cases in existence, and 9,376 fatalities, an unusual **slowing of Internet communications was noticed** – and progressed. An estimated 46,000 Floridians are trapped in shelters both due to fuel scarcities and fears about returning to the center of the epidemic in south Florida. Frightened public health officials are collaborating with local law enforcement to try and reduce further spread. The National Center for Communications of the Department of Homeland Security announced late in the evening the discovery of what appeared to be a **state-sponsored computer malware affecting both residential and commercial routers.** In the early hours of Jan 31, **a second computer virus began shutting down cell tower**

backbone routers, and a **third malware was found in trunked public service repeater systems** in three Florida counties., spreading along service systems to additional counties by the hour.

With the Northeast, West and Northwest still relatively unaffected by the avian influenza virus (named the "Quail Flu") but **beginning to lose internet and cell connectivity**, a Presidential announcement indicated the FBI had implicated a state-sponsored dual attack, but the sponsor was not fully determined. All armed forces were moved to active state, ships and submarines were sent to sea and the nation moved closer to martial law as food supplies dwindled in southeastern states, riots and arson were being reported throughout the media who were still on the air. As communications systems went offline, both fuel and frightened workers became more scarce, resulting in more media outlets going dark.

ALACHUA COUNTY

With shelters full and overflowing in Alachua County, internet systems slowing to a crawl and trunked radio systems inoperative, all possible County and City employees have been pressed into service to provide essential services, including ambulance service for new Quail Flue cases, fire service for the arson-induced conflagrations, efforts to quell riots in several neighborhoods and to provide communications between public safety units, authorities, shelters, and feeding centers.

It is now the morning of February 2nd, and skies are darkening rapidly to the northwest, the wind is picking up, and the barometer is dropping. The Alachua County Incident Commander is located at the EOC, reachable only by WINLINK (address **ALCTY-IC@WINLINK.ORG**) due to staffing shortages. *He has requested copies of any emails sent.* The State Emergency Department has deployed emergency communications assets, including amateur radio operators, reporting to a **SWIC at the State EOC (FLSWIC@WINLINK.ORG)** point of contact. He directs the **FLORIDA AMATEUR RADIO POINT OF CONTACT (FARPOC) Karl Martin (KG4HBN)** who (with a small team of assistants) is managing deployments of amateur radio operators across multiple counties, and currently based at a Staging Base at Santa Fe College with Amateur radio facilities and (for now) slow but working Internet.

Teams are being organized to reach **shelter #10**, 4 miles west of the Staging Base, from which a VHF Command Net has been running for the last 24 hours with exhausted staff, and **shelter #26** some miles to the east, from which there has been no word for 8 hours.

The FARPOC "Florida Amateur Radio Point of Contact" Management Group at work

Exercise ViralDuo After Action Report

2 INITIAL TASKS

The hierarchical organization of the Exercise was designed to replicate what was created by the Florida Dept of Emergency Management during Hurricane Michael, but replacing multiple deployed amateur radio single resource units, with two large Strike Teams

    Florida SWIC
    |
    Florida Amateur Radio Point of Contact ("FARPOC")
    |
    Strike Team 1   &   Strike Team 2

- SWIC participants drove to an on campus travel trailer whose antenna complement for vhf/hf had been set up at daybreak.

- FARPOC participants set up in the Conference main meeting room.

- Participants on Strike Team One and Strike Team Two had to get in vehicles and travel without cell phones to their assigned locations:

DEPLOYMENT INFORMATION FOR STRIKE TEAMS ONE AND TWO

(A) Paper Road Map is included with this Briefing Document. *Cell phones are out / no useful GPS*
Suggest your team utilize radio communications to facilitate vehicle travel to assigned sites as per ICS-205.   BE SAFE

(B) A special map of how TEAM ONE is to drive at Gordon's House (to avoid damaging sprinklers or getting stuck in a plowed garden) is also included. Review this!

B. STRIKE TEAM ONE – Gordon's House, NW 41st Avenue, Newberry, FL [It is NOT in Newberry, it is just 5 miles west of the Conference] Gate code is

#				

Drive onto the grass NOT hitting the orange-flag marked sprinklers and to the Solar-Panel festooned POLE BARN for cover. **SEE SITE MAP ON LAST PAGE Antenna Support:** You can use the nearby OAK TREE for antenna support. Do not attach antennas to the Pole Barn (to avoid possible damage to the Solar Panels from RF). Tractor and other items may be in the yard to facilitate tieing off ropes; free to use. KX4Z RMS Gateway is set to only 10Watts – please use only 10W when trying to contact it or on its frequencies to avoid possible damage. Guest house doors are unlocked to provide you with restroom (attached to bedroom) You can set up your equipment under the pole barn to be out of sun/rain. Avoid driving through the tilled garden (you will get stuck) Your Maidenhead Locator is: EL89rq

C. STRIKE TEAM TWO – Northside Park, Gainesville 5701 NW 34th Blvd, Gainesville, FL 32653; Park in paved public parking lot. There are 2 covered Pavilions, and bathrooms. Walmart is across the street. This is a FRISBEE PARK. Be careful when installing antennas, wires etc to avoid risks to bystanders; mark all wires/cables to prevent accidental bystander contact.

If any state or local unit should happen to arrive – **you are free to take advantage of any of their assets** with their permission to assist in your antenna deployment efforts!

Exercise ViralDuo After Action Report

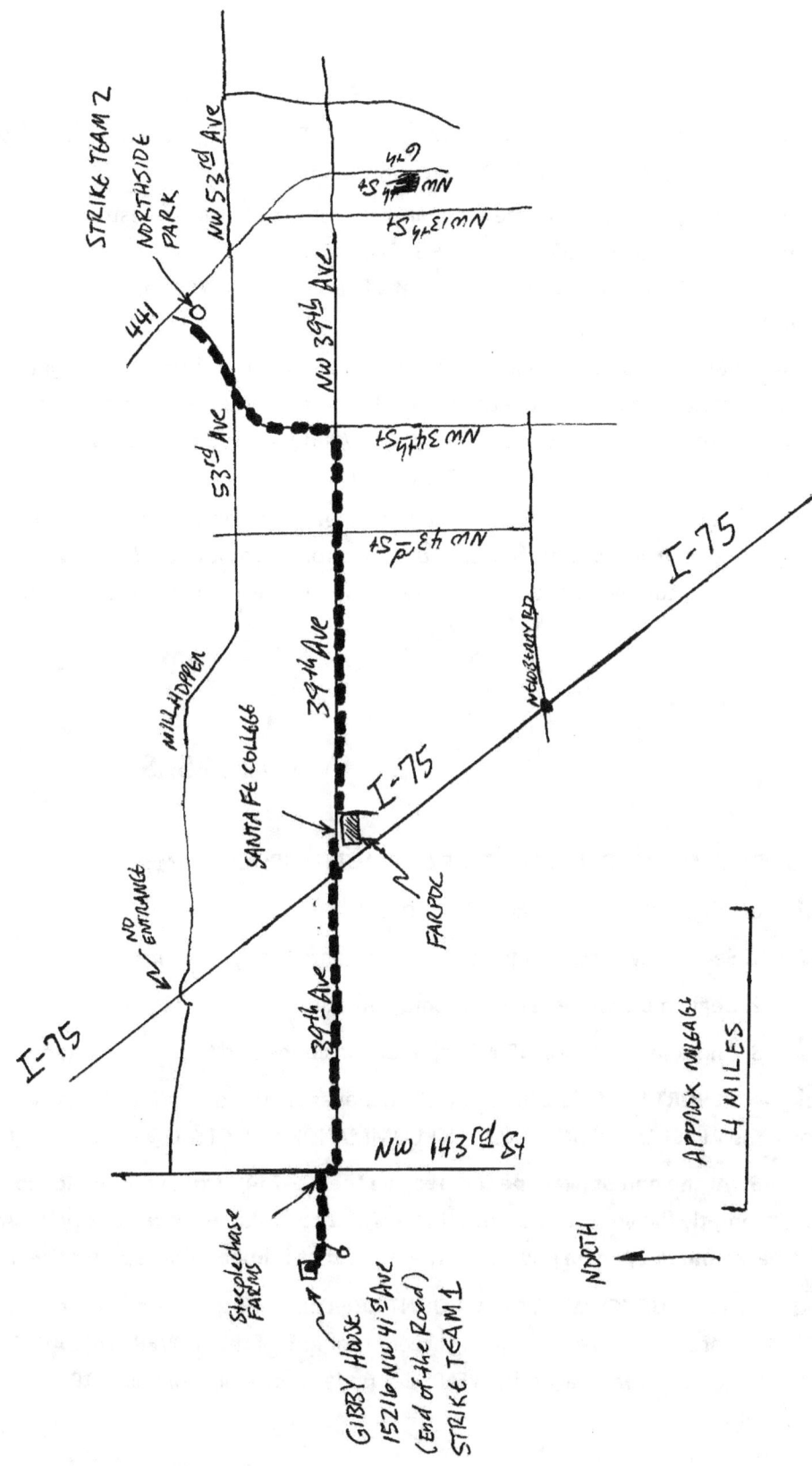

THE RULES OF THE GAME

1. You can use any system indicated on the ICS-205 or 205A unless it has been declared "failed" etc. You can use ANY frequency to do SIMPLEX communications.

2. Please do NOT use any of the other vast repeater and other infrastructure in the Gainesville area that are NOT on the ICS-205 or 205A. E.G., the SARNET is not allowed, private repeaters not listed are not allowed. If you BROUGHT a deployable repeater, that's FANTASTIC and YES you can use it (but we may "fail" it at some point).

3. The internet and other afflictions in this exercise apply to broad geographical areas – but if you are able to get a ham *more than one state away* to help you-- you can get them to send email for you, take phone calls for you – whatever! Pass messages by VOICE, CW, smoke-signals – whatever it takes. If you can do digital, GREAT, if you can't – GET IT DONE SOME OTHER WAY!

4. If you can bring up a Section or Area traffic net of any kind at all – t*hen any message that you send them, and they pass from ONE of their members to another (so it got relayed twice)* (simulating relaying out of the disaster area) can then be forwarded by ANY MEANS including internet, phone call, email, whatever

INITIAL TASKS

FLORIDA AMATEUR RADIO POINT OF CONTACT (FARPOC) -- Mission #0999

- ☐ 0. Sign in on your ICS-211 at your location.
- ☐ 1. **Set up** any missing antennas. Use any antennas you desire.
- ☐ 2. **Begin monitoring** assigned contact frequencies as per ICS-205A
- ☐ 3. **Check into** COMMAND NET and remain connected to that net whenever possible.
- ☐ 4. **REPORT** Send ICS-214 to the SWIC managing this effort (FLSWIC@WINLINK.ORG) and as a courtesy, copy the ALACHUA COUNTY INCIDENT COMMANDER (ALCTY-IC@WINLINK.ORG) who has requested this.
- ☐ 5. At the end of every period, receive the ICS-214's from your deployed units, and summarize situations and promptly file your ICS-214 with the SWIC & copy the Alachua County IC. Attempt to get this done before the next time to open an envelope arrives. If that is impossible, summarize and file as soon as practicable.
- ☐ 6. **EMERGENCY WINLINK ACCOUNT CREATION**. If there is one or more members of your team without WINLINK accounts, create a working account for one of them OVER THE RADIO. If everyone already has an account, Create a Tactical Address FLPOC-1 under the callsign of your FARPOC. If you need assistance, contact the SWIC

STRIKE TEAM ONE -- MISSION NUMBER 0123 BEFORE LEAVING THIS FACILITY DO ITEM (1)

Exercise ViralDuo After Action Report

☐ 1. Choose your STRIKE TEAM Leader and SHELTER MANAGER -- record your team's monitored callsign (i.e., winlink address) of that Team Leader on ICS-205A with the AUXCOMM LEADER before leaving this facility

☐ 2. Proceed to Shelter #10 location shown in SKETCHES above – note cell phone navigation is declared inoperative due to cell tower failures; paper map is provided herein.   DRIVE CAREFULLY. SIGN-IN on the ICS-211 at your deployment location.   Meet the SHELTER MANAGER!

☐ 3. COMMAND NET: Upon arrival, One subunit is to Immediately Stand up the Command Net ("Alachua County Simulated Emergency Net") on VHF voice as per ICS-205 attached. Establish contact and maintain continually with STRIKE TEAM TWO and the FARPOC LEADER

☐ 4. ANTENNAS: At arrival, expeditiously erect any needed antennae for VHF/UHF/HF communications and structure your sub-units YOU ARE ALLOWED TO USE THE OAK TREE NORTH EAST OF THE POLE BARN   DO NOT ATTACH ANTENNAS TO THE POLE BARN. You may use items in the Yard for rope connections, etc.

☐ 5. EMERGENCY WINLINK ACCOUNT CREATION.   If there is one or more members of your team without WINLINK accounts, create a working account for one of them OVER THE RADIO.   If everyone has an account, Create a Tactical Address FLTEAM-1 under the callsign of your Strike Team Leader. If you need assistance, contact the FARPOC..

☐ 6. STATUS REPORT: As soon as possible after arrival, send status report (ICS-214 preferred) to FARPOC, and additionally as a courtesy to the ALACHUA COUNTY IC as he requested this. (email: ALCTY-IC@WINLINK.ORG) The Internet is out.   Use any technique you can to get this done.

☐ 7. STAY CONNECTED: Continually monitor ALL frequencies (or nets) assigned to you per ICS-205 A, check for WINLINK traffic addressed to your STRIKE TEAM LEADER at least every 30 minutes, using any appropriate Gateway.

☐ 8. Create the ability to transmit on the lower end of the 160meter band, in preparation for possible utilization by Incident Commander. Document how you accomplished this task on ICS-214

☐ 9.. EXPECT: Expect new briefings / instructions at 1030 and 1130 and 1210, via the INJECT envelopes and/ or radio

☐ 10. REPORT: As soon as possible after arrival, send status report (ICS-214 preferred)   to FARPOC, and additionally if possible, as a courtesy to the Alachua County Incident Commander as he requested copies. (email: ALCTY-IC@WINLINK.ORG)   **Additionally repeat this 5-15 minutes before the end of each operational period.**   If you are not able to get your arrival ICS-214 out in timely fashion, just roll it into your first end-of-operational period report.

11. Record all traffic of significance (in either direction) on one or more ICS-309 forms. Retain this information until the end of the Exercise and submit by radio before leaving.

☐ 12. INJECTS: Open the INJECT Envelopes at the times printed on the outside of the envelopes and follow the instructions contained therein

☐ 13. STAY PUT: Except for emergency situations, do not take any other actions without discussion with the FARPOC or his delegated representative.

The Assignment for Strike Team Two was similar except that they were to join into the Command net.

ORGANIZATION:

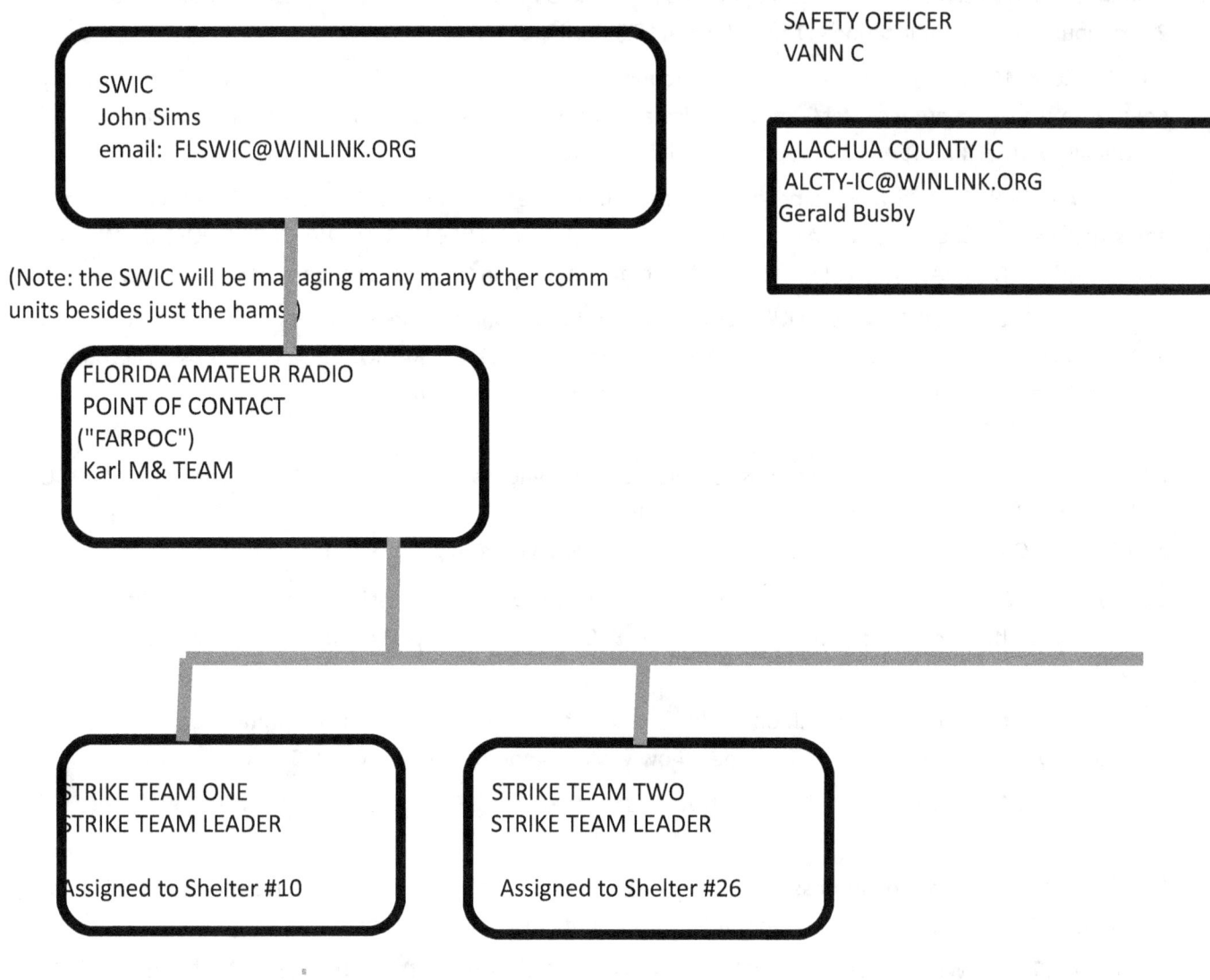

NOTE: there may be many other units in the field and other resources. Take advantage of them when appropriate. SHELTER MANAGERS come under the LOGISTICS UNIT of the local County, and are not depicted on this chart.

3 EXERCISE INJECTS

(Material taken from the Master Scenario Event List)

Homemade 2 meter antenna constructed on the fly to overcome one of the Injects

Event #	Event Time	Event Description	Responsible Controller	Recipient Player(s)	Expected Outcome of Player Action
	900	Start of Exercise (StartEx)			
01	900	Exercise Briefing and Division into ❏ teams and ❏ SHELTER MANAGERS ❏ OBTAIN CALLSIGNS AND WINLINK ADDRESSES FOR BOTH STRIKE TEAM LEADERS ❏ **Have everyone add those to their ICS-205A** ❏ **Paper Maps**	Gordon Gibby	All	Split group into 2 teams + FARPOC LEADER (team) Obtain call signs and addresses for both comm Strike Teams, and for Shelter Managers (phone #)
02	0930	❏ Deployment	ICS-201	All	Safely reach deployment locations (expect 15 minutes or more for travel)

Event #	Event Time	Event Description	Responsible Controller	Recipient Player(s)	Expected Outcome of Player Action
03	*0930ff*	❑ Set up Command NET (VHF primarily)	ICS-201	Strike Team One	Make connection with the other participant teams
04	*0930ff*	❑ Check into COMMAND NET	ICS-201	Strike Team Two FARPOC LEADER	Make connection with the command net
05	*0930ff*	❑ Set up all Antennas	ICS-201	Strike Team One Strike Team Two FARPOC LDR	Establish solid antennas for both VHF and HF
06	*0930ff*	❑Create WINLINK accounts using radio only	iCS-201	Strike team One Strike team Two FARPOC	Successfully create a new WINLINK account -- If they don't know how, SWIC can send them information
07	*0930ff*	❑Send Status Report to FARPOC LEADER and ALACHUA COUNTY IC	ICS-201	❑Strike Team One ❑Strike Team Two	Successfully send ICS214 Successfully deal with WINLINK Tactical addresses
08	*0930ff*	Monitor all assigned frequencies	ICS-201	Strike Team One Strike Team Two FARPOC	Monitor the frequencies assigned in ICS-205A (ICS-205 gives many frequencies but only those in the 205A must be actively monitored)

Event #	Event Time	Event Description	Responsible Controller	Recipient Player(s)	Expected Outcome of Player Action
09	0930ff	Record ICS-214 and ICS-309	ICS-201	Strike Team One Strike Team Two	Keep contemporaneous records
10	0930	Send report to Alachua Incident Commander and to the FARPOC Leader about progress	ICS-201	FARPOC	Identical reports should show up in the Winlink Tactical addresses for these two fictional players.
11	0945 1000 1015 1030	❏ Incident Commander (ALCTY-IC@WINLINK.ORG) check all incoming email		(SWIC)	Find ICS-214's from Strike Team ONE and Strike Team TWO and review their progress against expected items; find reports from FARPOC leader. NOTE TIMES FOUND
		1030 INJECT ENVELOPES			
12					
13	1030	FARPOC Leader has lost Internet communications	(envelope)	FARPOC LDR	Move to radios for communications
14	1030	Repeaters 146.82 and 146.985 become non functional and ❏ **Strike Team** One disappears from them (Exercise monitor will have to notify people on those repeaters)	(envelope)	Strike Team One FARPOC	(Also notify Neighborhood hams) (Note that COMM 2 is not notified of this; they will have to figure it out on their own) Move net to a different frequency - W4DFU may not work, so may require relays, or give up and go to HF.
15	1030	❏ Strike Team TWO – no 80 meter antenna	(envelope)	Strike Team Two	Move to 40 meters?

Event #	Event Time	Event Description	Responsible Controller	Recipient Player(s)	Expected Outcome of Player Action
16	1030	☐ **SWIC** Notify FARPOC LEADER that MEDICAL TEAM AND RUMOR CONTROL is available on 1840 LSB and ask that all teams be notified. ☐ **Exercise Controllers must begin to monitor 1840 LSB for medical calls / RUMOR CONTROL CALLS** ☐ **SWIC** Request updated ICS-214 from FARPOC Leader ☐ **SWIC** Ask FARPOC to have COMM ONE begin to monitor UTAC41 (453.46250	**SWIC (via winlink email)**	FARPOC LDR	FARPOC to notify teams of MEDICAL TEAM. FARPOC to send in ICS-214 FARPOC to notify Comm one of monitoring UTAC41
18	1030	☐ Formal Message (shelter status report) to transact from COMM ONE to FARPOC and request forwarding to SWIC	(envelope)	COMM ONE	(they have norovirus)
19	1030	FARPOC LEADER requested to forward message to state	COMM ONE request	FARPOC	
	1030	Unexpected Outside Volunteer W4DNA has radiogram delivered to FARPOC LEADER	W4DNA	FARPOC	Message is to ask FL STATE EOC TO CONTACT SOUTH CAROLINA AUXCOMM on 3.950 ASAP.
20	1045	Outside participant (Rebecca W.) contacts COMM TWO to ask them to provide communications for Shelter #35 @ 1000 N. Main Street - On their UHF repeater frequency	(outside volunteer)	COMM TWO	COMM TWO will initiate requests up the chain to find out what to do.
		1130 INJECT ENVELOPES			

Exercise ViralDuo After Action Report

Event #	Event Time	Event Description	Responsible Controller	Recipient Player(s)	Expected Outcome of Player Action
21	1130	❏ The 146.820 Repeater is working again	(envelope)	COMM ONE COMM TWO FARPOC LEADER	They may reconstitute the net there.
22	1130	❏ **FARPOC** is notified that meningitis is moving through shelters and RIFAMPIN is requested to be given	(envelope)	FARPOC	notify all shelters
23	1130	❏ **FARPOC** is notified that a software patch has been created and work to receive it on WINLINK	(envelope)	FARPOC	download "program"
24	1130	❏ COMM ONE loses all VHF Antennas has to replace them	(envelope0	COMM ONE	build small VHF antenna from parts supplied
25	1130	❏ COMM ONE requested to send iC-214 to FARPOC and either copy or have it forwarded to ALACHUA IC	(envelope)	COMM ONE	send ICS-214
26	1130	❏ **STRIKE TEAM ONE** to notify FARPOC leader of their loss of VHF antennas	(envelope)	Strike Team ONE	Use HF to reach FARPOC
27	1130	❏ **STRIKE TEAM** ONE to notify FARPOC Leader of any medications needed	(envelope)	Strike Team ONE	when notified of the rifampin request, request rifampin
28	1130	❏ **STRIKE TEAM** TWO generators dead but 80 meter antenna is now working again	(envelope)	Strike Team TWO	switch to batteries
29	1130	❏ Notify Net Control and FARPOC that there is a shelter resident with a fever and a stiff or achy neck.	(envelope)	SHELTER MANAGER TEAM TWO	Notify Net control station and FARPOC

Event #	Event Time	Event Description	Responsible Controller	Recipient Player(s)	Expected Outcome of Player Action
30	1130	❑ Formal message for STRIKE TEAM TWO to send to both Alachua FARPOC and STATE SWIC	(envelope)	SHELTER MANAGER TEAM TWO	send message by any formal means
		❑ **Rebecca W. contacts Strike Team one and tells them the tigers are free and devouring people**	RADIO REBECCA W	STRIKE TEAM ONE	
31	1140	❑ "Program" sent to FARPOC – must send email	SWIC	FARPOC	forward to all comm Strike Teams
		1210 INJECT			
32	1210	❑ Exercise is completed – send in ICS-214 / ICS-309 to FARPOC leader and to the ALACHUA IC, pack up and head back	(envelope)	COMM ONE COMM TWO FARPOC	send in reports and pack up and go home.

4 RESULTS

SAFETY
The Exercise was carried out without any known injuries or serious events. The Safety Officer reported minor issues with better marking of wires and ropes.

HOTWASH EVALUATION
A hurried hotwash was carried out (not enough time for a thorough analysis had been provided in the conference timeline) and based on verbal questioning of Team Leaders compared to known tasks, it appeared that roughly 40% of the approximately 64 assigned total tasks had been completed. However, team leaders may have answered conservatively, as their teams were LARGE and they may have been unable to assess every activity. **Lesson for Next Time: Provide adequate time for Discussion!**

WRITTEN PARTICIPANT EVALUATIONS
The participants were ***very enthusiastic*** about the exercise, based on written evaluatons returned.

Perception of difficulty:
0% thought the exercise was way too difficult;
32% thought it was difficult;
57% thought it was perfect;
11% thought it was easy; and
0% thought it was way too easy

Satisfaction with exercise:
0% found very poor,
0% poor;
11% neutral;
40% good satisfaction,
50% very good satisfaction

MESSAGE TRANSFER

Data capture of how many and which kinds of messages were transacted is incomplete, basically for two reasons:

1. I failed to allow enough TIME in the hotwash session for people to gather all their information and present an accurate view of exactly what they got done, and
2. Filling in ICS-309's was new to many people, and using the automated one inside WINLINK was new to almost all of us -- so people were not familiar with how to configure the report to capture the right days, times, boxes ("inbox" "outbox" "sent" etc -- capturing INBOX and SENT for the right time periods was the most useful)

As a result, the actual number of messages sent/received is almost certainly GREATER than the number reported here, just due to difficulties capturing the data. Where it was obvious that a message had to have been received (despite a missing report) I have entered a notation of "complementary message" for the presumed transation.

Although "origination" credit was not given, credit in this table is done similarly to National Traffic System counts for individuals; one sending and one receiving of the same messages results in a count of 2. (Nets count only 1 for a message transacted.)

TRAFFIC COUNTS	INFORMAL	FORMAL	
VOICE	18	12*	Total Voice 30
DIGITAL	0	48	Total Digital 48
Total	18 Informal	60 Formal	78 Grand Total

* One message notated on Team 1 ICS-309 does not indicate the mode used for transfer but appears to be a formal voice message.

Total Message Count (Send + Receive = 2 points)	78
Percent by Voice	38%
Percent by Digital	62%
Percent transacted as formal	77%
Percent Formal accomplished digitally	80%
Percent Formal accomplished by voice	20%

ANALYSIS OF MESSAGES MOVED

Review of messages received by the SWIC post shows an interesting trend: messages that were non-routine were moved reasonably well "up the chain" while routine (but just as easily important) reporting tasks were left undone.

Consider these messages received by the FLSWIC:

WINLINK TIME	SENT BY	CONTENT OF MESSAGE
1048	KF4VDF (TEAM 1)	ICS-213 formal message giving a shelter status report from Shelter #10 (this was a printed out full message in the Shelter Manager's envelope, which also included the message on a thumb drive.
1057	KM4JTE (TEAM 1)	Simple email informing FL SWIC that he was doing HF WINLINK
1144	KG4HVN	SWIC and ALCTY-IC were copied on a message from the FARPOC to both of the active winlinkers on Team 1, advising them to carry out a task to monitor on UTAC 41 -- the Formal ICS213 sent to the FARPOC was forwarded to TEAM 1 and copied back to the FLSWIC. This resulted from an INJECT by radio email sent to the FARPOC
1148	KG4HBN (FARPOC)	The FLSWIC received a message from the FARPOC advising that local medical personnel have detected meningitis in some shelters, and providing the correct dose of medication for victims. This communication resulted from an ENVELOPE INJECT opened by the FARPOC at 1130 -- and was therefore sent to the FLSWIC (and hopefully to the teams as well)
1156	KG4HBN	Email message notifying the FL SWIC that Shelter #26 has one resident with a fever and painful neck, and requesting assistance. This was a great example of an ENVELOPE INJECT to the SHELTER MANAGER of Shelter #26, which evidently was passed to the STRIKE TEAM 2, and from there to the FARPOC and from the FARPOC to the FLSWIC -- excellent work of moving critical information up the chain.
1200	KG4HBN	Email message regarding a report of a large tiger sighted south of "stable chase farms" reported at 11:38, observed by binoculars. This message was the result of an outside DISTRACTOR who reached TEAM ONE by radio and attempted to scare them with this RUMOR. The message went up the chain properly to the FARPOC and from there to the FLSWIC. Another example of an inject being moved rapidly and appropriately
1202	K4MVR (Strike Team 1)	An ICS213 requesting insulin for 3 individuals. This was apparently a spontaneously created Inject by the associated SHELTER MANAGER of Shelter #10 (who was empowered to do exactly such messages) -- possibly in response to an ENVELOPE INJECT to that Shelter manager at 1130 instructing them to ask for any medications for their shelter residents that they needed. Another example of a forwarded request. In this case it is unclear if it went also to the FARPOC.

| 1213 | KG4HBN (FARPOC) | Two different messages acknowledging receipt of an EXERCISE PRIORITY email generated by an OUTSIDE INJECTOR who injected a message that the FLSWIC should contact South Carolina AUXCOMM on 3.950 MHz LSB. This message was properly forwarded up the chain, acknowledged, and then the acknowledgments seemed to be acknowledged. |

What this seems to demonstrate is a well-oiled communication chain (with a few technical glitches elsewhere discussed) that is SPRING-LOADED to pass injects and other information up and down the chain! Bravo! What is missing however, was the new requirement listed and explained multiple times to send activity reports at the end of each operational period. Speculation: That could be because this was a "new idea" to send in operational period Activity Reports.

Note: See below for a compilation of message transaction details.

ANTENNAS AND SITUATIONS

TEAM	ANTENNA	COMMENTS
FL SWIC	VHF: end-fed vertical built into fiberglass mast; 25 feet VHF: mag mount ad-hoc placed on steel heater on top of travel trailer HF: non-resonant inverted off center fed homemade Windom fed to MFJ Intellituner HF: spare zip wire antenna that stretched far too much	Enormous interference (S9+20dB) from Champion 3400 watt inverter generator was recognized halfway through exercise; rendered 80/40 useless until corrected. Good digital connections on 20 meters even with generator going. Reached FARPOC on 40 meters later in the exercise to hunt for the HF net. VHF digital connections easy with 1 watt to NF4RC-3 (1/2 mile) VHF monitored team communications
FARPOC	Difficulties using the inverted vee non-resonant due to lack of Balun and tuner issues.	
STRIKE TEAM ONE	Virgil Team 1 was unable to hit any local stations, was able to hit Naples. Off center fed dipole john got it over the oak tree with his "dog toy" device. Multiple VHF and ad-hoc VHF antennas.	Comment: "Team went like ants to tasks" "Had there been a smaller team it would have been important to have it understood who was bringing what equipment." "Bob was busy tapping out ICS-213's on a computer"
STRIKE TEAM TWO	ST2 used the MARC tower as an elevated	Amateur operator reportedly wanted

			platform for the HF antenna. They opted not to put the VHF/UHF antennae on the tower. Unfortunately we were not able to park – the trailer close to the shelter due to vehicle access restrictions and the fact that the ground was very wet and soft. We set up in the parking lot, just under 200' from the shelter. The HF team had issues with tuning the antenna on 40m. It worked great on 20m and 80m but was useless on 40m even with the tuner.	to test out the performance of the new HF antenna.

COMPILATION OF MESSAGES TRANSACTED (BASED ON ICS-309)

A compilation of the messages transferred is as follows (and likely incomplete as requested forms were not easily located) Times are local times. ***This is a very large amount of traffic transferred!***

RCV   Message was documented as received
SEND/SENT   Message was documented as sent
Complementary Message Although not documented, message was almost certainly
           handled by counter-party.

STRIKE TEAM 1 K4MVR	STRIKE TEAM2 KN4MQQ	FARPOC KG4HBN	SWIC @WINLINK	ALCTY IC @WINLINK
			VHF **FORMAL WINLINK** 0919 KX4Z RCV TEST MESSAGE form KK4ECR [checking winlink accounts]	
			VHF **FORMAL WINLINK** 0937 KX4Z RCV message from KZ8Q from outside	
			VHF **FORMAL WINLINK** 1025 KX4Z RCV test message from FARPOC	
		-	**1030 (approx) FORMAL WINLINK**	

			SWIC INJECTS -- sent 7 messages to FARPOC.	
			VHF **FORMAL WINLINK** 1040 KX4Z SEND reply to KZ8Q Re; test message	
1048 KF4DVF **FORMAL WINLINK** SENT complete ICS213 shelter report & med request to FARPOC and ALCTY-IC		1048 **FORMAL WINLINK** RCV complete ICS-213 from KF4DVF (Team 1)		1048 **FORMAL WINLINK** RCV 213 copy from KF4DVF (Team 1)
1050 (NC) complementary message (Apparently received formally. No data on further forwarding.)	1050 FORMAL VOICE Formal Team2-213-2 1108 From KM4HCN TO "NC" -- ICS213 announcing 21 personnel on site. TX by KI4QBZ			
1054 (NC) complementary message NC replied no such shelter, but authorized dispatch (Reply appears informal)	1054 **FORMAL VOICE** Team2-213-1 I1105 From KM4HCN – other shelter needing communications – TO "NC"	(Cannot find a record of FARPOC being notified of this)	(	
1057 **FORMAL WINLINK** KM4JTE (TEAM 1) sent a test message to explain they were doing HF Winlink			1057 VHF **FORMAL WINLINK** (KX4Z RCV blank subj from KM4JTE (Team1)	
1102 (unknown mode, presumed voice) KK4DWE sent message requesting medical assistance to FARPOC			(cannot find this receipt on FARPOC records)	

			1113 VHF **FORMAL WINLINK** RCV FW Ex P from FARPOC	
			1117 VHF **FORMAL WINLINK** RCV This is a test from FARPOC	1117 HF **FORMAL WINLINK** RCV This is a test from FARPOC
(NC) complementary message	1119 VOICE Informal RCV from "NC' that there is no shelter #35			
	1120 VOICE Informal SENT (KM4HCN) to K9PDL – request info on comm request			
(NC) complementary message	1123 VOICE (informal) SENT (KM4HCN) to NC – 80 m antenna down.			
(NC) complementary message	1124 VOICE (informal) RCV from NC "dispatch from here" (?)			
(NC) complementary message	1127 VOICE (informal) SENT to NC dispatching to (unreadable)			
(NC) **FORMAL VOICE** complementary message	1131 **FORMAL VOICE** File: Team2-213-3.jpg SENT to NC "out of gas" going to batteries			
			1130 (approx) **FORMAL WINLINK** **INJECT** -- 2 messages to FARPOC	

		1133 **FORMAL WINLINK** RCV receipt from FLSWIC	1133 **FORMAL WINLINK** SEND RE: this is a test to FARPOC	
		1134 **FORMAL WINLINK** RCV ACK from FLSWIC	1134 **FORMAL WINLINK** SEND ACK Exercise P to FARPOC	
		1134 **FORMAL WINLINK** RCV 214 from K4MVR	1134 **FORMAL WINLINK** SEND Re: Exerciser P to FARPOC	
NC **FORMAL VOICE** (complementary message)	1137/1140 **FORMAL VOICE** SENT to NC (Informal?) Shelter mgr requests meds for patients. Appears to be file Team2-213.4			
NC (complementary message)	1141 VOICE (informal) RCV from NC Rumor control / Med Help on 1840			
NC (FORMAL VOICE) (complementary message)	1143 **FORMAL VOICE** File Team2-213-7.jpg 1143 (informal) RCV from "NC" primary repeater back on line			
		1144 SENT (RELAYED) **FORMAL WINLINK** 213 to advertised calls of Team 1 and 2 to monitor UTAC 41	1144 **FORMAL WINLINK** RCV UTAC41 note from FARPOC	1144 **FORMAL WINLINK** RCV UTAC41 note from FARPOC
NC (complementary message) **FORMAL** (No data that forwarded to FARPOC.)	VOICE 1146 **FORMAL** (Image: Team2-ICS-213-6.jpg) to NC; KN4KJB indicates wildlife risk- tigers observed			

Exercise ViralDuo After Action Report

	by binoculars loose			
		1148 **FORMAL WINLINK** SEND medical situation to FLSWIC	1148 **FORMAL WINLINK** RCV medical situation from KG4HBN	
NC (complementary message)	1154 VOICE (informal) RCV from "NC" "clarify Stable"			
		1156 **FORMAL WINLINK** SEND medical request to FLSWIC	1156 **FORMAL WINLINK** RCV medical request from KG4HBN	
NC (complementary message)	1200 VOICE (informal) SENT to "NC" 80 m antenna down	1200 **FORMAL WINLINK** SEND wildlife report to ALCTY-IC		1200 **FORMAL WINLINK** RCV wildlife report from KG4HBN
1202 SENT ICS- **FORMAL WINLINK** 213 requesting insulin for 3 persons, TO KK4ECR, ALCTY-IC, FLSWIC, FARPOC		1202 **FORMAL WINLINK** RCV 213 from K4MVR	1202 RCV ICS-213 **FORMAL WINLINK** requesting insulin from TEAM1	1202 RCV ICS-213 **FORMAL WINLINK** for insulin from TEAM 1
		1204 **FORMAL WINLINK** RCV ac of med request from FLSWIC		
			1210 ? VHF **FORMAL WINLINK** KX4Z send RE: VIRAL DUO to TEAM ONE	
NC (complementary message)	1211 VOICE (informal) RCV from "NC" Exercise over	1211 **FORMAL WINLINK** RCV ack of wildlife from ALCTY-IC		1211 SEND Re: Wildlife to FARPOC
		1213 **FORMAL WINLINK** SEND ACK of Exercise P NFL Sect Coord to FLSWIC	? VHF 1213 **FORMAL WINLINK** RCV ACK RE;: Exercise P from FARPOC	
		1213	?VHF	

| | | **FORMAL WINLINK** SEND RE: Exercise P NFL Coord | 1213 **FORMAL WINLINK** RCV Re: Exercise P from FARPOC | |
| | Note that Team 2 did not succeed with WINLINK. | | | |

Team 2 at Northside Park
All deployment locations had at least rudimentary shelter and restroom facilities.

5 EXERCISE CATEGORIZATION

Exercise Name	Viral Duo
Exercise Dates	02/02/19
Scope	This exercise is a full-radio, deployment exercise, planned for 3 hours at 4 different locations in Alachua County, Florida.. Exercise play is limited to radio communications.
Mission Area(s)	Response
Core Capabilities	Function in an ICS Framework; Create Antennas In a Devastated Deployment Location; Provide Electrical Power For Radios Independent of Commercial Utilities; Transact Multiple Types of Information By Radio
Objectives	Ability to read and understand ICS-201, 205, 205A; Ability to create and transmit ICS-214 and 309 forms; Properly respond to out of structure requests; properly respond to rumors; Deploying HF antenna; Deploying VHF antenna; Operate radios throughout the exercise without utility power; Operating within an HF SSB net; Operating within a VHF FM net; Obtaining WINLINK authorization; Discovering gateway frequencies; Connecting to Winlink gateways; Transacting email in the WINLINK system; Utilizing WINLINK tactical addresses; Programming VHF receivers; Transacting voice message traffic
Threat or Hazard	Novel respiratory virus with high fatality rate plus state-actor computer virus.
Scenario	Novel respiratory virus spreading rapidly through hundreds of miles, killing large numbers, destroying commerce and closing of states; complicated by the spread of a new computer malware that slows Internet and commercial communications (including trunked radio systems) to a crawl.
Sponsor	Santa Fe Amateur Radio Society and North Florida Amateur Radio Club
Participating Organizations	59 participants primarily from Florida, involved at their own expense; some involvement by County and Florida State assets (MARC Unit).
Point of Contact	Gordon L. Gibby MD KX4Z NCS521; docvacuumtubes@gmail.com

Team 1 furious digital entry.

6 ANALYSIS OF CORE CAPABILITIES

Aligning exercise objectives and core capabilities provides a consistent taxonomy for evaluation that transcends individual exercises to support preparedness reporting and trend analysis. Table 1 includes the exercise objectives, aligned core capabilities, and performance ratings for each core capability as observed during the exercise and determined by the evaluation team.

Objective	Core Capability	Performed without Challenges (P)	Performed with Some Challenges (S)	Performed with Major Challenges (M)	Unable to be Performed (U)
Ability to read and understand ICS-201, ICS-205, ICS-205A	Function in an ICS Framework		S		
Ability to create and transmit (by some means) ICS-214 and ICS-309 forms	Function in an ICS Framework			M	
Properly respond to out of structure requests	Function in an ICS Framework	P			
Properly respond to rumors	Function in an ICS Framework	P			
Deploying HF Antenna	Create Antennas in a devastated deployment location		S		
Deploying VHF Antenna	Create Antennas in a devastated deployment location	P			

Objective	Core Capability	Performed without Challenges (P)	Performed with Some Challenges (S)	Performed with Major Challenges (M)	Unable to be Performed (U)
Operate radios throughout the exercise without utility power	Provide electrical power for radios independent of commercial Utilities		S		
Operating within an HF SSB net	Transact multiple types of information by radio				U
Operating within a VHF FM net	Transact multiple types of information by radio	P			
Obtaining WINLINK authorization	Transact multiple types of information by radio				U
Discovering gateway frequencies	Transact multiple types of information by radio		S		
Connecting to winlink gateways	Transact multiple types of information by radio		S		
Transacting email in the WINLINK system	Transact multiple types of information by radio		S		
Utilizing WINLINK tactical addresses	Transact multiple types of information by radio	P			

Objective	Core Capability	Performed without Challenges (P)	Performed with Some Challenges (S)	Performed with Major Challenges (M)	Unable to be Performed (U)
Programming VHF receivers	Transact multiple types of information by radio	**P**			
Transacting voice message traffic	Transact multiple types of information by radio		**S**		

Table 1. Summary of Core Capability Performance

Ratings Definitions:

Performed without Challenges (P): The targets and critical tasks associated with the core capability were completed in a manner that achieved the objective(s) and did not negatively impact the performance of other activities. Performance of this activity did not contribute to additional health and/or safety risks for the public or for emergency workers, and it was conducted in accordance with applicable plans, policies, procedures, regulations, and laws.

Performed with Some Challenges (S): The targets and critical tasks associated with the core capability were completed in a manner that achieved the objective(s) and did not negatively impact the performance of other activities. Performance of this activity did not contribute to additional health and/or safety risks for the public or for emergency workers, and it was conducted in accordance with applicable plans, policies, procedures, regulations, and laws. However, opportunities to enhance effectiveness and/or efficiency were identified.

Performed with Major Challenges (M): The targets and critical tasks associated with the core capability were completed in a manner that achieved the objective(s), but some or all of the following were observed: demonstrated performance had a negative impact on the performance of other activities; contributed to additional health and/or safety risks for the public or for emergency workers; and/or was not conducted in accordance with applicable plans, policies, procedures, regulations, and laws.

Unable to be Performed (U): The targets and critical tasks associated with the core capability were not performed in a manner that achieved the objective(s).

The following sections provide an overview of the performance related to each exercise objective and associated core capability, highlighting strengths and areas for improvement.

Objective: Ability to read and understand ICS-201, ICS-205, ICS-205A (S)

The strengths and areas for improvement for each core capability aligned to this objective are described in this section.

Core Capability: Function in an ICS Framework

Strengths

The partial capability level can be attributed to the following strengths:

Strength 1: Previous experience of many of the participants with ICS-based exercises.

Strength 2: Participants generally succeeded well in finding frequencies, vehicular deployment, setting up command net, and related tasks which are similar to common experiences in Field Day or other group activities.

Areas for Improvement

The following areas require improvement to achieve the full capability level:

Area for Improvement 1: Carefully observing and completing assigned tasks in ICS-201 document (or equivalent).

Reference: http://wake.nc.auxcomm.us/wp-content/uploads/Example-ICS214.pdf

Analysis: Teams got a huge amount RIGHT!! Teams apparently dd not send in ICS-214 activity reports at the end of each operational Period despite this being clearly tasked in the ICS-201, and in envelopes. This was true at Strike Team 1, Strike Team 2, and FARPOC Team. There was apparently sufficient manpower and time in all teams, but for many, the concepts of sending in activity reports were new ideas and there was a significant lack of familiarity with ICS forms, particularly the ICS-214 and ICS-309 despite these being mentioned in pre-conference training. Additionally, the timeline of the Conference did not allow enough time for all participants to carefully read through the Exercise Materials. Although materials were available on the conference WIFI server, there wasn't specific time set apart for careful review. Suggest allowing more time and more carefully going through the Exercise objectives and tasks before deployment.

Objective: Ability to create and transmit (by some means) ICS-214 and ICS-309 forms (M)

The strengths and areas for improvement for each core capability aligned to this objective are described in this section.

CORE CAPABILITY: Function in an ICS Framework

Strengths

The partial capability level can be attributed to the following strengths:

Strength 1: A small portion of the participants were aware of the ability of WINLINK to automatically create an ICS-309.

Strength 2: A moderate portion of the participants were previously familiar with an Activity Report because of experiences during Hurricane Michael.

Areas for Improvement

The following areas require improvement to achieve the full capability level:

Area for Improvement 1: Familiarity with ICS-309 and ICS-214 forms.

Area for Improvement 2: Familiarity with WINLINK auto-creation of ICS-309 forms.

Area for Improvement 3: Practice sending ICS forms by voice as well as other techniques (packet, WINLINK, FLDIGI)

Reference:

Analysis: During the exercise, inject events and forms were well-handled, while there were essentially zero ICS-214 forms sent, which was surprising. Strike teams were not submitting requested Activity Reports and the FARPOC was not submitting either. These forms can be submitted by a plethora of techniques, such as
- voice (by simply reading them) packet – either by copy and paste, or by
- YAPP (available in EASYTERM by UZ7HO)
- NBEMS, using FLDIGI (or even by broadcast modes if the Signal to Noise ratio is good
- WINLINK, either via CMS or peer to peer

Participants utilized WINLINK (and possibly voice) well to transact urgent (INJECT) messages -- so there was some facility with communication. The problem was more with establishing a new habit of sending routine activity reports every operational period.

Objective: Properly respond to out of structure requests (P)

The strengths and areas for improvement for each core capability aligned to this objective are described in this section.

CORE CAPABILITY: Function in an ICS Framework

Strengths

The high capability level can be attributed to the following strengths:

Strength 1: Participants were familiar with working in a structured environment from previous net participation.

Areas for Improvement

The following areas require improvement to achieve the full capability level:

Area for Improvement 1: None

Reference:

Analysis: Participants sent word of out-of-structure requests to net control and/or other supervisors to obtain suggestions for best response. This was an adequate and reasonable response.

Objective: Properly respond to rumors (P)

The strengths and areas for improvement for each core capability aligned to this objective are described in this section.

CORE CAPABILITY: Function within an ICS Framework

Strengths

The high capability level can be attributed to the following strengths:

Strength 1: Participants were familiar with working in a structured environment from previous net participation.

Areas for Improvement

The following areas require improvement to achieve the full capability level:

Area for Improvement 1: None

Reference:

Analysis: Participants sent word of rumors to net control and/or other supervisors to obtain suggestions for best response. This was an adequate and reasonable response.

Objective: Deploying HF Antennas (S)

The strengths and areas for improvement for each core capability aligned to this objective are described in this section.

CORE CAPABILITY: Create Antennas in a devastated deployment location

Strengths

The partial capability level can be attributed to the following strengths:

Strength 1: Participants appeared to have significant skills at deploying lines over trees.

Strength 2: Participants quickly adjusted to working with available State assets such as the MARC trailer.

Areas for Improvement

The following areas require improvement to achieve the full capability level:

Area for Improvement 1: Familiarity with never-before-tested commercial antennas was lacking in Team 2.

Area for Improvement 2; Availability of suitable Balun(s) and wide-capability tuners was somewhat scarce in FARPOC

Reference: http://audiosystemsgroup.com/RFI-Ham.pdf https://www.qsl.net/nf4rc/BalunHowTo.pdf https://www.qsl.net/nf4rc/BalunPart1.pdf https://qsl.net/nf4rc/BalunPart2.pdf

Analysis: The HF net never quite appeared to take place. Unclear how much of that relates to antennas versus other skills, or understanding of the goals. SWIC was able to reach the FARPOC on 40 meters (from 1 mile away), recording a modest signal from the FARPOC. No other stations were ever heard on any HF voice frequency. SWIC was easily able to transact HF digital email on multiple bands with relatively high speed digital throughput. Reports from Team 1 location indicated that antenna was deployed and re-deployed in accordance with the task requirements. Report from Team 2 location indicated that a new commercial antenna was unable to be made workable on 40 meters, with participants reverting to a suggested issue by a previous speaker in the Conference. FARPOC participants were stymied by a balanced feed line non-resonant antenna of adequate height due to non-availability of a Balun or wide-ranging antenna tuner (their available tuner perhaps did not have the capability to match SWRs > 3;1?). The same antenna was used very effectively the next day using a MFJ Intellituner to make multiple WINLINK contacts in state after state with excellent signals. As many amateurs today live in urban environments, familiarity with HF antennas, Smith Charts, unusual transmission lines and antennas is declining. Well trained emergency participants and leaders may need to apply special efforts to gain the skills and assets to bring about full HF proficiency in difficult environments.

Objective: Deploying VHF Antennas (P)

The strengths and areas for improvement for each core capability aligned to this objective are described in this section.

CORE CAPABILITY: Create Antennas in a devastated deployment location

Strengths

The high capability level can be attributed to the following strengths:

Strength 1: Most participants had significant experience with VHF antennas.

Strength 2: A surprisingly large number of participants had experience with constructing a VHF antenna on the fly and knew how to reach a resonant length.

Areas for Improvement

The following areas require improvement to achieve the full capability level:

Area for Improvement 1: None

Reference: http://www.hamuniverse.com/2metergp.html   (many other possibilities as well)

Analysis: Very strong performance on deploying VHF antennas and also on constructing ad-hoc VHF antennas by Team 1. Team 2 worked well with the MARC unit to get their VHF antenna very high.

Objective: Operate Radios Throughout the Exercise Without Utility Power (S)

The strengths and areas for improvement for each core capability aligned to this objective are described in this section.

CORE CAPABILITY: Provide electrical power for radios independent of commercial Utilities

Strengths

The partial capability level can be attributed to the following strengths:

Strength 1: Participants had abundant sources of alternative power, even the SWIC with their difficulties with a generator, had two backup batteries and a backup inverter.

Areas for Improvement

The following areas require improvement to achieve the full capability level:

Area for Improvement 1: Find alternative power source for the Travel Trailer utilized by the SWIC, as the Champion 3400 watt generator was found to create insurmountable RF hash on 80 & 40 meters. The 2kw sine wave inverter, or 12V DC sources work fine for the radios, but do not suffice for Air Conditioning if required. NOTE: See later Appendix -- this problem was SOLVED.

Reference:

Analysis: Most teams appeared to have little difficulty powering their radios without utility power. Excellent skills and assets. However, unexpected insurmountable RF hash noise was noted on 80 and 40 meters in SWIC HF gear, after an hour of frustrating HF work on those bands – and tracked to the Champion 3400 watt inverter generator. Moving to a 12V storage battery to power the radio resulted in background noise reduction from S7 to S0. Connections on HF bands below 20 meters was then far easier. Even with the generator running, a high speed digital connection to N5TW was conducted on 20 meters using the 25 foot fiberglass mast non-resonant inverted vee antenna. The Champion generator made the same amount of RF has whether in econo mode or normal mode. See later appendix for extensive tests conducted to solve this issue.

Objective: Operate Within a SSB Net (U)

The strengths and areas for improvement for each core capability aligned to this objective are described in this section.

CORE CAPABILITY: Transact multiple types of information by radio

Strengths

The low demonstrated capability level can be attributed to the following strengths:

Strength 1: Some familiarity from more-experienced participants, including that gained during Hurricane Michael.

Areas for Improvement

The following areas require improvement to achieve the full capability level:

Area for Improvement 1: Actually putting together SSB net in field deployed position.

Reference:

Analysis: Unable to find any evidence of existence of an HF net during the exercise. Modest signal from FARPOC position noted. Unclear of any other successful HF Signals. Team 2 leader reported they never got 80 meters working... HF antennas are more unwieldy and as more amateurs live in restricted suburban environments, these antennas are possessed by fewer amateurs. ARES nets tend to be on VHF (and pass smaller amounts of formal traffic) while NTS and RRI nets tend to be on HF, pass more traffic, and have lower overlap with ARES members. Thus a concerted effort is required to maintain adequate penetration levels of HF assets and skills, sufficient to maintain HF SSB nets. It may be wise for emergency leaders to place special emphasis on developing these competencies.

Objective: Operate Within a VHF FM Net (P)

The strengths and areas for improvement for each core capability aligned to this objective are described in this section.

CORE CAPABILITY: Transact multiple types of information by radio

Strengths

The high capability level can be attributed to the following strengths:

Strength 1: Large amount of expertise at VHF nets.

Areas for Improvement

The following areas require improvement to achieve the full capability level:

Area for Improvement 1: Passing formal traffic

Reference:

Analysis: The VHF command net was quickly constituted. By report, when the primary repeater was "downed" the net migrated to a backup repeater on the ICS-205.

Objective: Obtain WINLINK Authorization (U)

The strengths and areas for improvement for each core capability aligned to this objective are described in this section.

CORE CAPABILITY: Transact multiple types of information by radio

Areas for Improvement

The following areas require improvement to achieve the full capability level:

Area for Improvement 1: Learning how to accomplish this task.

Reference: Chapter in Textbook given to participants. Also: https://qsl.net/nf4rc/2019/SettingUpWINLINKWithoutInternet.pdf

Analysis: This was a surprising failure. Neither Team 1, Team 2 or FARPOC completed this task (creating a winlink authorization completely by radio), despite explicit explanation in pre-conference email as well as printed instructions in a textbook handed to every participant upon arrival. The procedure takes only 2 connections to WINLINK and can be completed in less than 5 minutes in a disaster field. <u>Leadership should work to have this practiced in the coming year</u>.

Objective: Discovering gateway frequencies (S)

The strengths and areas for improvement for each core capability aligned to this objective are described in this section.

CORE CAPABILITY: Transact multiple types of information by radio

Strengths

The fairly high capability level can be attributed to the following strengths:

Strength 1: FARPOC volunteer(s) appeared to be quite successful at this. Team participants also seemed to succeed; previous practice (possibly with the Florida Winlink Check In Net) is likely the cause.

Areas for Improvement

The following areas require improvement to achieve the full capability level:

Area for Improvement 1: One participant, when confronted with having to use a different callsign, was stymied by obtaining MPS and gateway information over the radio. Team 2 was uanble to get WINLINK functioning.

Reference:

Analysis: Most teams did well at this. Review of how to accomplish over radio in a disaster theater far from familiar gateways may be indicated. With recent upgrades in WINLINK EXPRESS software, a new set of MPS stations and Gateways can be obtained over radio and a new propagation table very quickly created

Objective: Connecting to winlink gateways (S)

The strengths and areas for improvement for each core capability aligned to this objective are described in this section.

CORE CAPABILITY: Transact multiple types of information by radio

Strengths

The high capability level can be attributed to the following strengths:

Strength 1: Multiple participants succeeded well at this, transferring dozens of formal traffic.

Areas for Improvement

The following areas require improvement to achieve the full capability level:

Area for Improvement 1: Team 2 Leader reports their team never got WINLINK working --practice at this advised.

Area for Improvement 2: Participants would be well advised to obtain higher speed modes, including VARA and PACTOR for true disaster preparation.

Reference: http://arrl-nfl.org/wp-content/uploads/2014/02/NFLWinlinkPage-Speed-Advantage-of-Digital.pdf

Analysis: This was an area of strength in the participants of 3 of 4 teams, with urgent INJECT material well communicated, using WINMOR and other soundcard modes. However Team 2 was unsuccessful. Effort needed there! *Additional Suggestion:* PACTOR speeds are often twice that of soundcard modes, and connections are often made with far less signal required. Serious emergency communicators may wish to move to this pre-eminent digital protocol while keeping their competency at sound card modes. Participation in SHARES digital network or the DTN digital network also require PACTOR.

Objective: Transacting email in the WINLINK system (S)

The strengths and areas for improvement for each core capability aligned to this objective are described in this section.

CORE CAPABILITY: Transact multiple types of information by radio

Strengths

The high capability level can be attributed to the following strengths:

Strength 1: Multiple participants had significant experience in WINLINK. Perhaps 5 had been regular participants in the Florida Winlink Check In net, which tests all nooks and crannies of WINLINK systems (well beyond the basic email) week after week. Experienced personnel were strategically utilized to make communications for multiple teams.

Areas for Improvement

The following areas require improvement to achieve the full capability level:

Area for Improvement 1: Some Participants were less familiar with sending FORMS

Area for Improvement 2: **Most** Participants were less familiar with sending ICS-309

Area for Improvement 3: One participant confused which mode was appropriate for 2 meter packet (and used WINMOR on a packet frequency for about 30 minutes)

Area for Improvement 4: One participant accidentally shifted audio tone frequencies for AX.25 resulting in a frustrating period of unsuccessful connections.

Area for Improvement 5: Team 2 was unable to make WINLINK contact.

Reference:

Analysis: Creating an ICS-309 takes less than 2 minutes on WINLINK but no participant transmitted one over radio. PACKET (AX.25) is the correct mode to send to connect to 2-meter packet gateways and sending WINMOR is both ineffective and reduces possible throughput on the frequency. Accidentally shifting the packet tones (which are different for HF than VHF) is easy to do – unless you have "locked the pointers" in UZ7HO soundmodem.com Inexperience with the software leads to such difficulties.

Digital signals definitely take more familiarity with systems and software than, say, sending CW. Accidentally changing packet audio frequencies is similar to accidentally setting Receiver Incremental Tuning – and both can lead to zero connections. However, since packet on VHF is a simplex technology, if the operator has the ability to monitor the frequency at the same time, other operators can come on FM voice and explain to them their problem. That was tried in both these accidental situations, but neither operator appeared to be able to hear the helpful corrective information. Adding a speaker to their system would be a good move.

The advantage gained for the additional complexity is vastly greater throughput, which is analogous to the advantage gained by switching from Amplitude Modulated double sideband signals to Single Side Band in the 1950's and 1960's – however, those modes now are very easy to use, while using digital modes still requires some "computer smarts" which is unfortunately not widespread in the amateur radio system at present.

Objective: Utilizing WINLINK tactical addresses (P)

The strengths and areas for improvement for each core capability aligned to this objective are described in this section.

CORE CAPABILITY: Transact multiple types of information by radio

Strengths

The high capability level can be attributed to the following strengths:

Strength 1: Users appeared unfazed by unusual winlink addresses and simply used them.

Areas for Improvement

The following areas require improvement to achieve the full capability level:

Area for Improvement 1: Teach skills of how to create winlink tactical addresses

Reference: https://www.winlink.org/content/tactical_addresses

Analysis: Two winlink tactical addresses were utilized extensively in this exercise – FLSWIC@WINLINK.ORG and ALCTY-IC@WINLINK.ORG. These were helpful in identifying the function of the position. Team 1 and Team 2 had no such tactical address so the group had to add their WINLINK addresses to their ICS-205A's, adding a possible point of failure, rather than having, say, TEAM1 and TEAM2 as WINLINK tactical addresses.

Leadership should know how to create winlink tactical addresses on the fly in a disaster theater and be able to coach others how to do this. In future exercises, this may allow the use of tactical addresses by all teams.

Objective: Programming VHF receivers (P)

The strengths and areas for improvement for each core capability aligned to this objective are described in this section.

CORE CAPABILITY: Transact multiple types of information by radio

Strengths

The high capability level can be attributed to the following strengths:

Strength 1: Participants appeared to have the skill for programming common VHF amateur radio transceivers to receive interoperability frequencies, and were also able to find the correct frequency.

Areas for Improvement

The following areas require improvement to achieve the full capability level:

Area for Improvement 1: Add the ability to program non-transmit (e.g., using CHIRP).

Reference: (Using CHIRP, simply set Duplex to "OFF" to prevent transmission on a stored channel.)

Analysis: This was a task from last year, when a number of participants were unable to complete it. It appears that skill was gained in this task in the intervening year. Adding the ability to turn off TRANSMITTING on an interoperability frequency would be a plus.

Objective: Transacting voice message traffic (S)

The strengths and areas for improvement for each core capability aligned to this objective are described in this section.

CORE CAPABILITY: Transact multiple types of information by radio

Strengths

The partial capability level can be attributed to the following strengths:

Strength 1: Participants appeared to be very proficient at transacting informal traffic.

Strength 2: There was also significant ability to transact formal traffic.

Areas for Improvement

The following areas require improvement to achieve the full capability level:

Area for Improvement 1: Gaining more ability in transacting formal traffic.

Reference:

Analysis: **Although last in this list of Objectives, the ability to pass formal traffic simply using a microphone is definitely NOT the least of the list!** The majority of formal traffic in this Exercise was transmitted using WINLINK, even though there are multiple other possibilities (voice, FLDIGI, YAP). Continued analysis of the results indicated significant usage of VOICE for passing formal traffic during this exercise -- EXCELLENT! The ability to pass traffic by voice, with appropriate procedural words to speed corrections, is essential to a competent emergency communicator. Our Conference included a large dose of teaching on this subject (two compete sessions). Further practice at this skill is likely needed on an ongoing basis to remember procedural words and VHF versus HF frequency utilization. .

7 IMPROVEMENT PLAN

Item	Core Capability	Area For Improvement	Corrective Action	Suggested Primary Organization	Outcome (FILL IN AS REPORTED)
1	Function in an ICS Framework	Carefully observing assigned tasks in an ICS-201 document (or equivalent)	Take more time to guide the assembled participants to view the assigned tasks, and go over them carefully.	Florida Amateur Radio Emergency Communications Conference	
2		Familiarity with ICS 309 and 214 forms	Go carefully over these forms prior to next exercise	Florida Amateur Radio Emergency Communications Conference	
3		Familiarity with WINLINK auto-creation of ICS-309 forms	Go carefully over this feature at next Conference. SEC's initiate training on this feature with EC's.	Florida Amateur Radio Emergency Communications Conference Section Emergency Coordinators:	
4		Practice sending ICS forms by voice as well as other means	Florida ARES nets to practice these forms in significant numbers	Three Florida ARRL Sections	
5	Create Antennas in a devastated deployment environment	Familiarity with personally owned antennas	Request participants to test their own antennas Better interchange between Section HF and VHF nets to	Florida Amateur Radio Emergency Communications Conference SEC's and STM's	

			improve HF capabilities of typical ARES members		
6		Availability of suitable Balun(s) and wide capability tuners in FARPOC	Procure suitable equipment[1]	NFL SEC; WCF SEC; Team members	
7		Alternative power source for Travel Trailer	Develop low-pass-filter for generator and/or shielding; or replace generator.	Travel Trailer Owner	**COMPLETED.** See the Appendix: Quieting the Inverter Generator
8	Transact multiple types of information by radio	SSB net in field deployed position	Repeat this task in local exercises during the next year, and repeat in next Conference	NFL County ARES groups and Florida Amateur Radio Emergency Conference	
9		Passing formal traffic	Capture data on formal traffic in County ARES nets and require significant practice.	SEC's	
10		Obtaining WINLINK Authorization	All NFL/WCF EC / AEC to obtain winlink authorization	SEC's	
11		Obtaining MPS/ gateway information over radio	All NFL/WCF EC / AEC to practice obtaining MPS/gateway information over radio	SEC's	
12		Obtain high speed protocols such as VARA and PACTOR	All NFL/WCF EC / AEC to have WINMOR and ARDOP protocols functioning; encourage VARA / PACTOR	SEC's	
13		Sending Forms	All NFL/WCF EC / AEC to participate in FORMS practice on	SEC's	

---

1 There are innumerable suitable Baluns and tuners. Some that can be recommended include: LDG **RBA-1:1 Balun** Current Balun and LDG **RBA-4:1 Voltage Balun** MFJ 918; MFJ 911; and the highly acclaimed MFJ 993B Intellituner.

			FL WINLINK Check in Net		
14		Sending ICS-309	All NFL/WCF EC / AEC to create ICS-309 and send	SEC's	
15		Correct modes	Training of ARES County groups on Winmor vs packet modes	SEC's	
16		Accidental audio tone shift	Training of EC / AEC on locking soundmodem pointers One week training on this topic amongst Florida WINLINK training net	SEC's Florida Winlink Check in Net	
17		WINLINK Tactical Addresses	Training of key EC / AEC in creating Tactical Addresses	NFL SEC; WCF SEC	
18		CHIRP non-transmit feature on interoperability frequencies	Demonstration of CHIRP to participants	Florida Amateur Radio Emergency Communications Conference	
19		Formal voice traffic transactions	(repeat from above) Capture data on formal traffic in County ARES nets and require significant practice.	SEC's	

APPENDIX: WRITTEN EVALUATION DOCUMENT

1. The level of difficulty of the exercise was (check one)

Way Too Difficult	Difficult	Perfect	Easy	Way Too Easy

2. The most difficult part of the exercise was (write in your answer)

3. The easiest part of the exercise was (write in your answer)

4. The worst feature of the Exercise was

5. The best feature of the Exercise was

Exercise ViralDuo After Action Report

APPENDIX: ICS-309 DOCUMENTS RECEIVED

FL SWIC TEAM

(Two computers in operation, accidentally resent some emails from larger computer)

Planned inject messages were pre-created on 1/23-1/27, saved in a "drafts" folder and moved into the Outbox and sent during the Exercise. The ICS-309 was auto-compiled by Winlink Express Software. Times listed for SENT messages are the CREATION time for the message, not the actual SEND time.

COMMUNICATIONS LOG			TASK # Gainesville Full Scale Exercise	DATE PREPARED: 2019-02-10 TIME PREPARED: 19:04
OPERATIONAL PERIOD # 2/2/2019 9:30 AM - 12:30			TASK NAME: Gainesville Full Scale Exercise (Big Computer)	
RADIO OPERATOR NAME: Gordon Gibby			STATION I.D. KX4Z	
LOG				
TIME	FROM	TO	SUBJECT	
2019-01-23 14:50	FLSWIC	KG4HBN kg4hbn@gmail.com	//WL2K 1030 FARPOC: RUMOR CONTROL	
2019-01-23 16:09	FLSWIC	KG4HBN kg4hbn@gmail.com	//WL2K 1030 FARPOC NEW ICS205 FOR 1030 OPERATIONAL PERIOD	
2019-01-24 13:27	FLSWIC	KG4HBN kg4hbn@gmail.com	//WL2K 1030 FARPOC: MED TEAM AVAILABLE	
2019-01-24 13:27	FLSWIC	KG4HBN	//WL2K 1030 FARPOC ICS205A CHANGES	
2019-01-24 13:27	FLSWIC	KG4HBN	//WL2K 1030 FARPOC SUBMIT ICS214	
2019-01-24 13:27	FLSWIC	KG4HBN	//WL2K 1130 FARPOC ROUTER REPAIR PROGRA	
2019-01-27 08:43	FLSWIC	KG4HBN	//WL2K 1030 FARPOC: UPDATE TO ICS 201 BRIEFING	
2019-01-27 17:43	FLSWIC	KG4HBN	//WL2K 1030 FARPOC STRIKE TEAM ONE UTAC41	
2019-02-02 09:19	KK4ECR	KX4Z	Re: //WL2K TEST MESSAGE	
2019-02-02 09:37	KZ8Q	KX4Z	//WL2K test priority message	
2019-02-02 10:25	KG4HBN	KX4Z	Re: //WL2K TEST MESSAGE	
2019-02-02 10:40	KX4Z	KZ8Q	Re: //WL2K test priority message	
2019-02-02 11:13	KG4HBN	FLSWIC	FW: //WL2K Exercise P NFL Section Coordinator	
2019-02-02 11:17	KG4HBN	FLSWIC ALCTY-IC	//WL2K This is a test	
2019-02-02 11:33	FLSWIC	KG4HBN	Re: //WL2K This is a test	
2019-02-02 11:34	FLSWIC	KG4HBN	ACK: FW: //WL2K Exercise P NFL Section Coordinator	
2019-02-02 11:34	FLSWIC	KG4HBN	Re: FW: //WL2K Exercise P NFL Section Coordinator	

Big Computer corrected Count: 17 ; Previous count: 9 (8 added)

Gordon Gibby KX4Z

COMMUNICATIONS LOG			TASK # Gainesville Full scale exercise -- SMALL COMPUTER	DATE PREPARED: 2019-02-10
				TIME PREPARED: 19:49
OPERATIONAL PERIOD # 2/2/2019 9-12:30			TASK NAME: GAINESVILLE FULL SCALE EXERCISE SMALL COMPUTER-2	
RADIO OPERATOR NAME: Gordon I. Gibby				STATION I.D. KX4Z
LOG				
TIME	FROM	TO	SUBJECT	
2019-01-24 13:27	FLSWIC	KG4HBN kg4hbn@gmail.com	//WL2K 1030 FARPOC: MED TEAM AVAILABLE	
2019-01-24 13:27	FLSWIC	KG4HBN	//WL2K 1130 FARPOC ROUTER REPAIR PROGRA	
2019-01-24 13:27	FLSWIC	KG4HBN	//WL2K 1030 FARPOC SUBMIT ICS214	
2019-01-24 13:27	FLSWIC	KG4HBN	//WL2K 1030 FARPOC ICS205A CHANGES	
2019-01-27 11:20	FLSWIC	KG4HBN	//WL2K 1130 FARPOC ICS 205 205A 201 CURRENT OPERATIONAL PERIOD	
2019-01-27 17:47	FLSWIC	KG4HBN	//WL2K 1030 FARPOC STRIKE TEAM ONE UTAC41	
2019-02-02 10:48	KF4DVF	ALCTY-IC	//WL2K ICS 213: Exercise Viralduo	
2019-02-02 10:57	KM4JTE	KX4Z	//WL2K	
2019-02-02 11:44	KG4HBN	ALCTY-IC FLSWIC	FW: //WL2K 1030 FARPOC STRIKE TEAM ONE UTAC41	
2019-02-02 11:48	KG4HBN	FLSWIC	//WL2K test message medical situation	
2019-02-02 11:56	KG4HBN	FLSWIC	//WL2K test message medical request	
2019-02-02 12:00	KG4HBN	ALCTY-IC	//WL2K test message report of wildlife	
2019-02-02 12:02	K4MVR	ALCTY-IC FLSWIC	//WL2K ICS 213: Viral Duo	
2019-02-02 12:04	FLSWIC	KG4HBN	Re: //WL2K test message medical request	
2019-02-02 12:10	KX4Z	K4MVR	Re: //WL2K ICS 213: Viral Duo	
2019-02-02 12:11	ALCTY-IC	KG4HBN	Re: //WL2K test message report of wildlife	
2019-02-02 12:13	KG4HBN	FLSWIC	ACK: Re: FW: //WL2K Exercise P NFL Section Coordinator	
2019-02-02 12:13	KG4HBN	FLSWIC	Re: FW: //WL2K Exercise P NFL Section Coordinator	
2019-02-02 12:26	FLSWIC	KG4HBN	ACK: Re: FW: //WL2K Exercise P NFL Section Coordinator	

SWIC Small Computer ICS-309 -- 19 transactions (previous count was: 12, so 7 more here).

OVERLAP: 6 messages are duplicates.
TOTAL SENT BY SWIC 17+19= 36; 6 of which are duplicates, so corrected count is 30.
Total added are 15 messages, of which 6 are dupes so 9 new messages total.

Exercise ViralDuo After Action Report

FARPOC

COMMUNICATIONS LOG	TASK # ICS 309 Symposium packet WINLINK	DATE PREPARED: 2019-02-03 TIME PREPARED: 15:27
OPERATIONAL PERIOD # 2 February 8:00-13:00		TASK NAME: ICS 309
RADIO OPERATOR NAME: Susan KG4VWI		STATION I.D. KG4HBN
LOG		

TIME	FROM	TO	SUBJECT
2019-02-02 10:48	KF4DVF	KG4HBN	//WL2K ICS 213: Exercise Viralduo
2019-02-02 11:33	FLSWIC	KG4HBN	Re: //WL2K This is a test
2019-02-02 11:34	FLSWIC	KG4HBN	ACK: FW: //WL2K Exercise P NFL Section Coordinator
2019-02-02 11:34	FLSWIC	KG4HBN	Re: FW: //WL2K Exercise P NFL Section Coordinator
2019-02-02 11:34	K4MVR	KG4HBN	//WL2K ICS214-Exercise Viralduo
2019-02-02 11:48	KG4HBN	FLSWIC	//WL2K test message medical situation
2019-02-02 11:56	KG4HBN	FLSWIC	//WL2K test message medical request
2019-02-02 12:00	KG4HBN	ALCTY-IC	//WL2K test message report of wildlife
2019-02-02 12:02	K4MVR	KG4HBN	//WL2K ICS 213: Viral Duo
2019-02-02 12:04	FLSWIC	KG4HBN	Re: //WL2K test message medical request
2019-02-02 12:11	ALCTY-IC	KG4HBN	Re: //WL2K test message report of wildlife
2019-02-02 12:13	KG4HBN	FLSWIC	ACK: Re: FW: //WL2K Exercise P NFL Section Coordinator
2019-02-02 12:13	KG4HBN	FLSWIC	Re: FW: //WL2K Exercise P NFL Section Coordinator

Gordon Gibby KX4Z

STRIKE TEAM 1

(Hand generated, two forms, using form from Blank Book)

COMM Log ICS 309-SCCo ARES/RACES	1. Incident Name and Activation Number: Viral Duo	2. Operational Period (Date/Time) From: To:
3. Radio Net Name (for NCOs) or Position/Tactical Call: Comm Net 3		4. Radio Operator (Name, Call Sign): Ron KK4DWE

5. COMMUNICATIONS LOG

Time (24:00)	FROM Call Sign/ID	Msg #	TO Call Sign/ID	Msg #	Message
	KK4DWE	1	FARPOC KH4ECR	1	Need Medical assistance to shelter, call 911
	"	"	FARROC	1	Shelter Status Report

COMM Log ICS 309-SCCo ARES/RACES	1. Incident Name and Activation Number: Viral Duo	2. Operational Period (Date/Time) From: 2FEB 1000 EST To:
3. Radio Net Name (for NCOs) or Position/Tactical Call: TEAM 1		4. Radio Operator (Name, Call Sign): KM4MVR

5. COMMUNICATIONS LOG

Local Time (24:00)	FROM Call Sign/ID	Msg #	TO Call Sign/ID	Msg #	Message
1050	KF4DWE	ICS213-10	FARPOC	1	ICS 213 Shelter Status Rpt
1102	KK4DWE		FAROC	2	Request medical asst.
1100	KM4JTE		GoBox		HF Winlink Tst

Team 1 ICS-214

Muster 214
Team 1

ACTIVITY LOG (ICS 214)

1. Incident Name:

2. Operational Period: Date From: 2/2/19 Date To: 2/2/19
Time From: 0930 Time To: 1230

7. Activity Log (continuation):

Date/Time	Notable Activities
02/02/0950	Left conf.
/1010	Arrived at Shelter A
/1015	Set up VHF station, monitor Command Nets
/1045	Changed to Net central Freq #3
/1105	Medical Situation 4 call 911
/1115	Monitoring of 458.7125 MHz — Tac #2
/10:30	3910 Radio ck w/ NAPLES [VQAllmond] N42U
1:19	VHF Packet winlink station oper (?)
1046	Contact HF and sent winlink msg
1140	N9ENE built 1/4λ VHF antenna - operational
1141	Winlink account established over radio
1144	WT4OD built 1/2λ dipole for 2m
1144	Medic unit on 1840.
1145	Weather High Winds overnight
1145	Primary repeater is backup
"	Remain on N 3rd repeater
1147	Replacement VHF ant operational
11:39	Report of large Tigers S of Steeplechase Farms @ 19:78 observed by binoculars

8. Prepared by: Name: Rob Downs **Position/Title:** Rad. **Signature:** /s/

ICS 214, Page 2 **Date/Time:**

STRIKE TEAM 2

(Hand generated on form from Blank Book)

Time (24:00)	FROM Call Sign/ID	Msg #	TO Call Sign/ID	Msg #	Message
1108	KM4HCN	1 - 21 people personnel app	NC		21 people personnel RJaynes NW shlt #35/main st, req. Comm
1115	"	2	NC		No #35 - correction
1119	NC		Team 2		(Team 2 simplex) - try to confirm
1120	KM4HCN		K9PDL		80m down
1123	KM4HCN		NC		Dispatch from there - KN9?
1124	NC		Tm 2		- we are sending GG - KN4BAA to #
1127	Tm 2		NC		- no gas opr, on Batt power
1131	Tm 2		NC		- SM learned 26 puls med asst (fever)
1137/1140	Tm 2		NC		184D-O - Hiwnts overnight, no rumor
1141	NC		Tm 2 (rtm)		Prim. rep online - remain on Tere.
1143	KK4ECR/NC		All st.		forw. msg KN4KTB eg wildlife - obs
1146/1149	T2		NC		Clarity STABLE
1154	NC		T2		Status - 80m down
1200	T2		NC		
1211	NC		T2		Exercise complete

Appendix E – ICS Form 213 – NIMS Message Format

GENERAL MESSAGE

0001

TO: STRIKE 2 LEADER POSITION: DAVE
FROM: KN4KJB POSITION:
SUBJECT: LARGE WILDLIFE DATE/TIME: Feb 2 11:39

MESSAGE:
Report large Tigers wondering around south of STABLE CHASE FARMS. REPORTED AT 11:28, OBSERVED BY BINOCULARS IN THAT AREA

DATE: 2 FEB TIME: 11:39 SIGNATURE/POSITION: Brian Shulther K4BJS Operator

RECEIVED FROM: KN4KJB DATE: 2 FEB 19 TIME: 11:39 MSG NUMBER: 0001
SENT TO: Net Control DATE: 2 Feb 19 TIME: 11:49

213 ICS 1/79 – Modified SCD RACES 5/08
IFES 1336

(REPRESENTATIVE ICS-213 TRANSACTED)

Exercise ViralDuo After Action Report

APPENDIX: QUIETING THE INVERTER GENERATOR

The inverter generator powering the SWIC Team radio travel trailer was discovered to be putting out enormous RFI on 80 meters. The Improvement Plan includes a task to solve that serious issue. This appendix reports the results of 3 hours of testing on solutions.

Equipment:

Champion 3400 Watt Dual Fuel Inverter Generator $948
Ref: https://www.amazon.com/Champion-3400-Watt-Portable-Inverter-Generator/dp/B01FAWMMEY

Xantrex ProWatt 2000 watt Sine Wave Inverter,
Model 806-1220 $368

Older 3.5 kW gas standard generators unknown

Older 10 kW Generac generator (2-cylinder Subaru engine) unknown

ICOM IC-725 Solid State Transceiver

MFJ Intellituner Model MFJ993B

Photo of Chicken Wire Shielding and RFI Filter
(neither of which is grounded to the generator frame in this photo.)

Frequency: 3.586 MHz;
130 foot Inverted Vee Off center fed antenna, from 25 foot fiberglass mast at rear of travel trailer.
CW signal sent in order to tune antenna to excellent SWR

Test	Situation	Measured Noise	Comment
1	Baseline. No generator, operating radio from 12 volt trailer battery	XX------------------------------------ S0 S1 S2 S3 S4 S5 S6 S7 S8 S9 S9+10 S9+20db	
2	Inverter generator running,, connected to trailer, radio from trailer battery (being charged)	XXXXXXXXXXXXXXXXXXXXXXXXXXXXXXXXXXXXXXXXXX--- S0 S1 S2 S3 S4 S5 S6 S7 S8 S9 S9+10 S9+20db	
3	Inverter generator running, connected to trailer with power RFI filter at generator, radio from trailer battery (being charged)	XXXXXXXXXXXXXXXXXXXXXXXXXXXXXXXXXXXX-------- S0 S1 S2 S3 S4 S5 S6 S7 S8 S9 S9+10 S9+20db	
4	Inverter generator merely running, near trailer, radio running from battery alone	XXXXXXXXXXXXXXXXXXXXXXXXXXXXXXXXXXXXXXXXX--- S0 S1 S2 S3 S4 S5 S6 S7 S8 S9 S9+10 S9+20db	Suggests significant radiated energy
5	Inverter generator going, powering trailer through power RFI filter	XXXXXXXXXXXXXXXXXXXXXXXXXXXXXXXX---------- S0 S1 S2 S3 S4 S5 S6 S7 S8 S9 S9+10 S9+20db	Econo mode no different.
6	Inverter generator going, NOT CONNECTED to trailer, radio off trailer battery alone	XXXXXXXXXX---------------------------------- S0 S1 S2 S3 S4 S5 S6 S7 S8 S9 S9+10 S9+20db	No wires to generator, big improvement
7	Inverter generator, RFI filter, plugged into trailer, but battery charger fuses pulled, radio from trailer battery alone	XXXXXXXXXXXXXXXXXXXXXXXXXXXXXX---------- S0 S1 S2 S3 S4 S5 S6 S7 S8 S9 S9+10 S9+20db	
8	Inverter generator running, 20 yards away, not connected, radio running off trailer battery	XXX-------------------------------------- S0 S1 S2 S3 S4 S5 S6 S7 S8 S9 S9+10 S9+20db	
9	Inverter generator running, RFI filter, 20 yards away, powering trailer battery, radio off trailer battery.	XXXXXXXXXXXXXXXXXXXXXXXXXX---------------- S0 S1 S2 S3 S4 S5 S6 S7 S8 S9 S9+10 S9+20db	
10	**No generator, operate from 120VAC generated by Xantrex 2kW sine wave inverter**	XX------------------------------------ S0 S1 S2 S3 S4 S5 S6 S7 S8 S9 S9+10 S9+20db	No discernible noise.
11	Conventional 3.5 kW generator 60 yards away, no filter.	XXXXXXXXXXXXX------------------------- S0 S1 S2 S3 S4 S5 S6 S7 S8 S9 S9+10 S9+20db	
12	Conventional 10 kW generator 60 yards away, no filter	XXXXXXXXXXXXXX------------------------ S0 S1 S2 S3 S4 S5 S6 S7 S8 S9 S9+10 S9+20db	
13	Repeat Baseline, operate off battery alone	XXXXXXXXXX---------------------------- S0 S1 S2 S3 S4 S5 S6 S7 S8 S9 S9+10 S9+20db	Variations in background noise with passing time.
14	Inverter generator used to		

	operate transformer based 12V charger to charge battery powering Xantrex 2KW inverter 120VAC.	XXXXXXXXXXXXXXXX------------------------ S0 S1 S2 S3 S4 S5 S6 S7 S8 S9 S9+10 S9+20db	
15	Inverter generator, 20 yards away, Chicken Wire around one axis of generator, radio off charging trailer battery	XXXXXXXXXXXXXXXXXXXXXXXX-------------- S0 S1 S2 S3 S4 S5 S6 S7 S8 S9 S9+10 S9+20db	Grounding radio to (-) makes no difference.
16	Inverter generator, 20 yards away, Chicken wire around one axis of generator, RFI filter on AC output of generator; operate radio from 120VAC	XXXXXXXXXXXXXXXXXXXXXXXXXX------------ S0 S1 S2 S3 S4 S5 S6 S7 S8 S9 S9+10 S9+20db	Grounding radio to ground of 120VAC makes no difference
17	Inverter generator, 20 yards away, **chicken wire around one axis of generator – connected by clip lead to ground chassis of generator**; operate radio from 120VAC	XXXXXXXXXXXXXX-------------------------- S0 S1 S2 S3 S4 S5 S6 S7 S8 S9 S9+10 S9+20db	Big improvement from grounding chicken wire cage.
18	**Inverter generator, 20 yards away, chicken wire around one axis of generator, connected to generator frame ground; RFI filter on 120VAC output, case of filter connected to chicken wire/frame ground; Radio operating off 120VAC**	BEST CASE INVERTER GENERATOR XXXXX------------------------------------ S0 S1 S2 S3 S4 S5 S6 S7 S8 S9 S9+10 S9+20db	**HUGE IMPROVE-MENT**
19	Same as 18 but remove RFI filter from generator output	XXXXXXXXXXXXXXXXXXXXXXXX-------------- S0 S1 S2 S3 S4 S5 S6 S7 S8 S9 S9+10 S9+20db	RFI filter is vital.
20	Same as 18 but remove chicken wire around generator	XXXXXXXXXXXXXXXXXXXXXXXX---------------- S0 S1 S2 S3 S4 S5 S6 S7 S8 S9 S9+10 S9+20db	Chicken wire is ALSO vital!
21	Same as 18 but run radio from 12 trailer battery being charged by generator	XXXXXXXXXXXXXXXXXXX--------------------- S0 S1 S2 S3 S4 S5 S6 S7 S8 S9 S9+10 S9+20db	12V trailer power being charged inferior to 120VAC power
	ADDITIONAL TESTS RUN ON 10KW CONVENTIONAL GENERAC GENERATOR APPROX 60 YARDS AWAY		
22	Baseline: powered by battery	XXX-------------------------------------- S0 S1 S2 S3 S4 S5 S6 S7 S8 S9 S9+10 S9+20db	
23	Trailer powered by 10kw generator with no alterations, radio running of 12V battery charged from trailer	XXXXXXXXXXXXXX-------------------------- S0 S1 S2 S3 S4 S5 S6 S7 S8 S9 S9+10 S9+20db	
24	Trailer powered by 10 kw	XXXXXXXXXXXX----------------------------	

	generator, no alterations, radio running from 110 VAC	S0 S1 S2 S3 S4 S5 S6 S7 S8 S9 S9+10 S9+20db	
25	10 kw Generator with RFI filter on output, grounded to frame	XXXXXXXXXXXX------------------------------ S0 S1 S2 S3 S4 S5 S6 S7 S8 S9 S9+10 S9+20db Generator appears to be radiating spark plug or other wide band noise. Chicken wire was not tried, but might help.	The same both DC and AC power of radio

FINAL SOLUTION:

If Air Conditioning or other large power draw is required, move inverter generator 20 yards at right angles to HF Antenna, wrap in chicken wire, ground chicken wire to frame of generator, Use 30A RFI filter on output of generator, and ground frame of RFI filter to generator frame. Expect negligible increase in background noise.

If Air Conditioning or other large power draw is NOT required operate radios from Xantrex 2kw sine wave inverter from battery sources and expect no interference.

If using a conventional generator -- suggest encasing generator in chicken wire and grounding chicken wire to frame, or taking other steps to reduce ignition wide band noise.

ABOUT THE CONFERENCE

The 2019 Amateur Radio Emergency Communications Conference was the 2nd Annual gathering hosted by the Santa Fe Amateur Radio Society and the North Florida Amateur Radio Club. Both conferences brought together scores of amateur radio operators at their own expense to train for emergency communications. We do not know of any other such hands-on, deployment exercise-based Conference for volunteer citizen amateur radio operators in Florida. While deeply connected to the American Radio Relay League, the Conference is also open to all amateur radio operators and all non-governmental organizations with emergency communications needs, and all city, county, state or federal authorities with an interest in emergency communications.

NORTH FLORIDA AMATEUR RADIO CLUB (NFARC) WEB PAGE
https://www.qsl.net/nf4rc/

The 2019 Conference was extremely well received, with scores (on a 1-5 scale) exceeding 4.5 in exit evaluations received. Over and over the participants cited the Exercise as one of the greatest portions of the conference, and also praised the many hands-on sessions that gave them direct connection to radio and electronic technology.

Some Texts published by Gordon L. Gibby on behalf, some on behalf of the North Florida Amateur Radio Club:

Amateur Radio Digital and Voice Emergency Communications: Build your community group's assets & expertise, 2nd Edition
https://www.amazon.com/Amateur-Radio-Digital-Emergency-Communications/dp/1548004340

Steinhatchee Storm: "How-To" Puerto-Rico Style Ham Radio Full Scale Exercise: Helping your volunteer ARES group carry out a Full Scale Exercise (Alachua ARES Exercises) (Volume 3)
https://www.amazon.com/Steinhatchee-Storm-How-Puerto-Rico-volunteer/dp/1978441509

2018 Amateur Radio Emergency Communications Symposium
https://www.amazon.com/Amateur-Radio-Emergency-Communications-Symposium-ebook/dp/B079JRYHHV

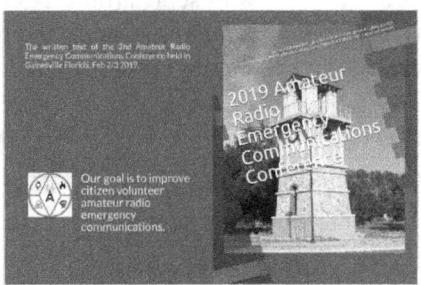

2019 Amateur Radio Emergency Communications Conference: North Florida Amateur Radio Club Santa Fe Amateur Radio Society (NFARC Conferences)
https://www.amazon.com/Amateur-Radio-Emergency-Communications-Conference/dp/1791865941

Below is the planned schedule for the Conference, to show the extremely diverse nature of the training offered.

Planned Schedule

	SATURDAY
0700	**REGISTRATION** **Please come early to REGISTER so we can start on time!**
0800	**Introduction** Gordon Gibby / Jeff Capehart
0815	**Volunteer Ham Radio Team Building that Maximizes all Volunteers** Joe Bassett - Confirmed (45 min)
0900	**Using ICS Documents to Manage Large Deployments** **& Introduction to the Full Scale Exercise / Team Splits** Gordon Gibby – Confirmed (15 / 15 minutes)
0930	**Begin Full Scale Exercise 3 Hours**

	Team 1 (Unit 1) Team 2 (Unit 2) Logistics Chief (Karl Martin & Deputies)		
1230	Reassemble from Full Scale Exercise Hot Wash Sessions by Teams / LUNCH (Moderator: Gordon Gibby)		
1315	How to Plan / Create / Carry Out Full Scale Exercises & BUILD YOUR GROUP Leland Gallup – Confirmed 30 Minutes		
1345	Lessons from Cascadia Rising & Oregon 2018 SET (15 minutes) Gordon Gibby – Confirmed		
1400	Section Emergency Coordinator's Message Karl Martin, Section Emergency Coordinator – Confirmed 30 Minutes		
	BEGIN SPLIT SESSIONS		
	R-1 Main Conference room	**Downstairs Dining R-106**	**Upstairs Dining R203**
1430 (45 min)	TALK 101 **Baptist Disaster Relief Services for Amateur Radio Volunteers** Marvin Corbin, Logistics/Field Missionary, Florid Baptist Disaster Relief Confirmed 25 registered	TALK 102 **WIFI Shelter Bulletin System To Keep Shelter Residents Informed** *Copies of the microSD for the Raspberry will be available at cost (est. $6) [If you bring a raspberry pi and a cheap wifi home router, you can turn it on immediately]* Gordon Gibby - Confirmed 20 registered	TALK 103 **Introduction to Publicity / PIO** Scott Roberts, Section PIO – confirmed 19 registered
1515 (45 min)	TALK 202 **Moving Traffic and Training Volunteers in ARES NETS** PART ONE (Joe Bassett) - confirmed Part ONE 30 registered	TALK 201 **MARC UNIT DEMONSTRATION** KEVIN RULAPAUGH depending on weather, may be outside at the trailer or inside in a room. confirmed Outside or in downstairs dining R106 ? 15 registered (talk changed)	TALK 203 **Computer & Internet Tips for EMCOMM** **PART ONE** Files – copying, directories Installing Applications Virus Prevention Jeff Capehart – confirmed 20 registered
1600 (45 min)	TALK 302 **Moving Traffic & Training Volunteers** **PART TWO**	TALK 301 **Hurricane Michael Experiences** James Lea – confirmed	TALK 303 **Computer & Internet Tips for EMCOMM**

	(Joe Bassett) - confirmed 37 registered	(Talk has changed so registration #s aren't accurate)	**PART TWO** Purchasing Computers Updates Advanced Skills Jeff Capehart – confirmed 18 registered
1645	WRAP UP OF DAY ONE Gordon Gibby Jeff Capehart		
5 PM	ADJOURN		

SUNDAY
DAY TWO

	R1 Main Conference Room	R106 Downstairs Dining (unless otherwise specified)	R203 Upstairs Dining (unless otherwise specified)
1200 (90 min)	TALK 403 **Repeater Controller for your own Voice Repeater – setting up the ICS-CTRL inexpensive repeater controller** 1200-1300 Gordon Gibby – confirmed ------------------------ 1300-1330 TALK 404 **Hands-on Solar Power Systems.** Gordon Gibby – confirmed 31 registered	Session 401 **Hands On Session – Power Pole connectors.** This session will show you how, and actually INSTALL power pole connectors wherever you need them. $1 each (+&-), bring your radios and power supplies, learn and add connectors. https://www.amazon.com/Didamx-Anderson-Powerpole-Disconnect-Connector/dp/B07D333JM2 Mentors signed up: Karl Martin – confirmed John Troupe- confirmed Alvin Osmena – confirmed 17 registered	Session 402 **ROOM HA-105** Hands-On Session – Wiring your radio for signalink / digital This session will actually wire your radio for digital. Preregistrants need to list their radio and whether they have an empty mic connector. Bring your actual radio and a connector for the microphone or accessory jack and we'll try to make the cables necessary for you at cost. Connections will be to the "Alachua County ARES Standard" Multiple mentors Susan Halbert - confirmed

			Mike Ridlon – confirmed 15 Registered
1330 (90 min)	Session 502 **Hands-on VHF/HF GO BOX BUILDING** First Come First Served / Build Go Boxes for your VHF or HF radio systems at cost. Mentors Alvin Osmena – Confirmed Stewart Reissener – Confirmed 25 registered – **will use the OUTSIDE TABLES ALSO**	Session 501 **ROOM HA-105** **AD HOC VHF ANTENNA BUILDING** (actually build antennas – at cost) Mentors; Susan Halbert – confirmed Alternate – Alvin confirmed **(Can move to HA105 if a lot of soldering is required)** 19 registered	TALK 503 **Tuning a Duplexer with a $110 Chinese Spectrum Analyzer** Gordon Gibby – confirmed 19 registered
1500 (90 min)	Session 601 **Hands-On WINLINK Training** Simultaneous tutors: Gordon Gibby – confirmed . (We are going to need more tutors!!!) 35 registered	Session 602 **Secrets to building rapidly deployable HF multiband antennas in hurricane devastated areas** Mentors: Leland Gallup Shannon Boal 20 registered	Session 603 **ROOM HA-105** **Solder Session Get Started on a $20 Soundcard digital System** **USE ROOM HA-105** You will get started on actually building your digital interface system – at cost. MENTORS Alvin Osmena – confirmed John Troupe – confirmed 5 Registered

1630 (30 min)	TALK 703 **Working Well with your EOC** Jeff Capehart – confirmed 31 registered	ALK 701 **Neighborhood Ham Watch** Gordon Gibby – Confirmed 17 registered	TALK 702 **Teaching Ham Radio Courses using ARRL Slides** Shannon Boal 15 registered
1700	Course Certificates & Graduation		

www.ingramcontent.com/pod-product-compliance
Lightning Source LLC
Chambersburg PA
CBHW081613220526

45468CB00010B/2862